FUNDAMENTALS OF SUPERCONDUCTIVITY

FUNDAMENTALS OF SUPERCONDUCTIVITY

Vladimir Z. Kresin
Lawrence Berkeley Laboratory
University of California
Berkeley, California

and

Stuart A. Wolf
Naval Research Laboratory
Washington, D.C.

PLENUM PRESS • NEW YORK AND LONDON

Library of Congress Cataloging-in-Publication Data

Kresin, Vladimir Z.
Fundamentals of superconductivity / Vladimir Z. Kresin and Stuart A. Wolf.
p. cm.
Includes bibliographical references and index.
ISBN 0-306-43474-1
1. Superconductivity. I. Wolf, Stuart A. II. Title.
QC611.92.K74 1990
537.6'23--dc20 90-42066
CIP

First Printing—September 1990
Second Printing—June 1992

ISBN 0-306-43474-1

A Division of Plenum Publishing Corporation
233 Spring Street, New York, N.Y. 10013

Printed in the United States of America

PREFACE

The recent discovery of high-temperature superconductivity has resulted in a remarkable growth in the amount of research and the number of researchers working in this exciting field. Superconductivity is not a new phenomenon: in 1991 it will be 80 years old. Even though it was the newer discoveries which motivated us to write this book, the book itself is mainly a description of the fundamentals of the phenomenon.

The book is written for a very broad audience, including students, engineers, teachers, scientists, and others who are interested in learning about this exciting frontier of science. We have focused on the qualitative aspects, so that the reader can develop a basic understanding of the fundamental physics without getting bogged down in the details. Because of this approach, our list of references is not comprehensive, and it is supplemented with a summary of additional reading consisting of monographs and selected review articles. (The articles we have referenced were either not reflected in the review articles on monographs or were milestones in the development of the field.) In addition, some of the sections which can be skipped during the first reading have been marked with asterisks (*).

Until recently, superconductivity was considered to belong to the field of low-temperature physics. This field was born, simultaneously with quantum physics, at the beginning of this century. Initially these two contemporaneous fields developed independently, but they soon became strongly coupled. One can understand superconductivity and related phenomena only through comprehension of the details of quantum physics. Superconductivity and superfluidity are the most remarkable manifestations of quantum regularity on a macroscopic scale. We therefore expect the reader to be acquainted with the basics of quantum theory.

One of us (VZK) is unable to overestimate the help provided by Vitaly Kresin in dealing with style, translation, and presentation of the science. We also are very grateful to Kenneth Schubach for his careful editing of the manuscript and many valuable suggestions.

Also, we are very grateful to Lilia Kresin and Iris Wolf. It is only because of their encouragement and sacrifice that this book has become a reality.

Vladimir Z. Kresin
Stuart A. Wolf

CONTENTS

1. A LITTLE HISTORY. SUPERCONDUCTING MATERIALS 1

The Breakthrough to Absolute Zero 1
The Discovery of Superconductivity 3
Superconducting Materials 5
Superconductors in a Magnetic Field 8
Thermal Properties 11
The Mystery of Superconductivity 12
The Isotope Effect 14

2. THE NATURE OF SUPERCONDUCTIVITY 17

Microscopic Theory 17
Finite Temperatures. The Critical Temperature 22

3. ENERGY GAP 27

Tunneling 27
Ultrasound Attenuation 32
Energy Gap Anisotropy 33
Superconductors with Overlapping Bands 35
Gapless Superconductivity 37

4. MACROSCOPIC QUANTIZATION 39

Quantization of Magnetic Flux 39
The Josephson Effect 42

5. THERMAL AND ELECTROMAGNETIC PROPERTIES 47

Heat Capacity 47
Thermal Conductivity of Superconductors 48
Anisotropic Thermal Effect 54
Electromagnetic Properties of Superconductors. Penetration Depth .. 55
The Ginzburg–Landau Theory 58
Critical Fields. The Mixed State 60
Hard Superconductors. Critical Current 64
Superconductors in a Variable Electromagnetic Field 65
Knight Shift 68

6. THE EFFECTS OF STRONG COUPLING. CRITICAL TEMPERATURE 69

Influence of the Phonon Spectrum. Eliashberg Equation 69
Main Relations of the Theory of Strong Coupling 71
The Function $g(\Omega) = \alpha^2(\Omega)F(\Omega)$. Tunneling Spectroscopy 73
Critical Temperature 76
Coupling Constant λ 79
Once More about the Isotope Effect 81

7. NON-PHONON MECHANISMS OF SUPERCONDUCTIVITY 85

Pairing Mechanisms 85
Coexistence of Phonon and Non-Phonon Mechanisms. Identification of Non-Phonon Mechanisms 90

8. SUPERCONDUCTING FILMS 93

Two-Dimensional Superconductivity 93
Granular Superconductivity 95
Localization 97
Proximity Effect. Induced Superconductivity 97
Preparation of Superconducting Films 101

9. SUPERCONDUCTING SYSTEMS.................... 109

A15 Materials.................................... 109
B1 Compounds...................................... 112
Organic Superconductors.......................... 112
Heavy Fermions................................... 115
Superconductivity in Semiconductors.............. 116
Oxides and Hydrides.............................. 118

10. THE SUPERCONDUCTING STATE IN NATURE....... 121

Superfluidity of Liquid Helium................... 121
The "Superconducting" State in Nuclei............ 132
Superconductivity and Astrophysics............... 133
Superconductivity and the Physics of Complex Molecules......... 137

11. MEASUREMENT TECHNIQUES.................... 141

Critical Temperature............................. 141
Critical Current Measurements.................... 145
Critical Magnetic Field Measurements............. 150

12. APPLICATIONS OF SUPERCONDUCTIVITY......... 155

Superconducting Electronics...................... 155
SQUIDs... 157
Superconducting Digital Electronics.............. 166
Power Applications of Superconductivity.......... 168
Power Transmission and Distribution.............. 175
High-Power RF Applications....................... 175

13. HIGH-T_c CUPRATES....................... 179

New Materials.................................... 179
Common Properties of the Cuprates................ 185
Preparation of the Cuprate Superconductors....... 190
Critical Current Densities....................... 191
Critical Magnetic Fields......................... 193
Status of Applications........................... 193

APPENDIXES **195**

A. Bound States in Two Dimensions 197
B. The Method of Elementary Excitations (Quasiparticles) 199
C. Electrons in Metals. Fermiology 201
D. Plasmons 205
E. What is Unique about the High-T_c Oxides? 209

SUGGESTED READINGS **223**

REFERENCES **225**

INDEX **229**

FUNDAMENTALS OF SUPERCONDUCTIVITY

1

A LITTLE HISTORY. SUPERCONDUCTING MATERIALS

THE BREAKTHROUGH TO ABSOLUTE ZERO

If the earth were taken into very cold regions, for instance, to those of Jupiter or Saturn, the water of our rivers and oceans would be changed into solid mountains. The air, or at least some of its constituents, would cease to remain an invisible gas and would turn into the liquid state. A transformation of this kind would thus produce new liquids of which we as yet have no idea.

—Antoine Lavoisier

At least until recently, superconductivity has been observed only at low temperatures, near absolute zero. For this reason, we would like to begin by describing the main achievements on the way to absolute zero.

The great French chemist Lavoisier spoke the words quoted above in the middle of the 18th century. During the 19th century, many gases were liquefied, but Lavoisier's dream came completely true only in 1908, when in the Leiden laboratory of the Dutch physicist Kamerlingh-Onnes the last remaining gas—helium—was finally turned into a liquid.

A very common method of liquefying gases is to compress them. This forces the molecules closer together, the role of cohesive forces increases, and the gas–liquid transition becomes possible. However, if the gas is at a temperature higher than the critical point, no pressure can turn it into a liquid. In this case, it is necessary to cool the gas before compressing it.

By 1870 a great number of liquefaction experiments had been carried out, and a large number of new liquids obtained. However, hydrogen, nitrogen, and oxygen showed no signs of turning into liquids. As a result, physicists began to think that these three were "permanent gases." Only in 1877, the French scientist Cailletet succeeded in obtaining liquid oxygen at a temperature of 90.2 K. Six years later, nitrogen was liquefied, at 77.4 K.

The attainment of liquid hydrogen was very complicated. It appears only at a temperature of 20.4 K. This problem was solved only in 1898 by Sir James Dewar, a professor at the Royal Institution in London. To store the liquid hydrogen, he invented a vacuum vessel which is used to this day and has been named after him. In solving the problem of liquefaction of hydrogen, Dewar had no doubts that his work was the last step on the way to the absolute zero, but he was mistaken. The boiling point of liquid helium turned out to be even lower.

Helium was discovered in 1869 in the spectrum of the solar corona. For a long time, nobody was able to find it on the earth. In 1895, Sir William Ramsay found it among gases released when certain minerals were heated. At the very end of the 19th century, it became clear that the boiling point of helium was even lower than that of hydrogen.

In the late 1890s, Kamerlingh-Onnes began his experiments aimed at liquefying helium. He succeeded in writing the final chapter in the history of the search for "new liquids." In his laboratory, on July 10, 1908, physicists from different countries, specially invited to observe the historic experiment, first saw liquid helium. The experiment began at six o'clock in the morning and lasted for 16 hours. During this time, Kamerlingh-Onnes and his assistants were in a state of tremendous tension (for several months afterward, Kamerlingh-Onnes could not continue his work in the lab because of extreme exhaustion). The experiment produced about 60 cm^3 of liquid helium. The transition temperature is just 4.2 K.

Liquid oxygen, nitrogen, and hydrogen were obtained by very clever but modest means. Kamerlingh-Onnes was the first to realize that the experimental physicist in the 20th century had to be a good engineer as well. He saw that the attack on helium required new, powerful technology. He founded a famous school of glassblowers and machinists and constructed a special refrigerator.

Nowadays, nobody would be surprised by this experimental approach. Everybody knows, for example, that modern accelerators are not just experimental apparatus, but are technically enormously complicated constructions. The same applies to modern radio telescopes, electron microscopes, high-pressure devices, and other equipment of modern physics. However, at the turn of the century Kamerlingh-Onnes stood out among many contemporary experimentalists who used simple, small-scale setups.

So, in 1908 the last natural gas—helium—was liquefied. Around that same time, Planck's work on thermal radiation and Einstein's work on the photoelectric effect laid the foundation of the quantum theory. At first, quantum physics and low-temperature physics developed independently. The future revealed the close ties between these two contemporary fields. Kamerlingh-Onnes' work played a large role in these developments. In his Nobel lecture (1913), expounding on his interest in low-temperature physics, Kamerlingh-Onnes said that research in this field allowed one to "contribute towards lifting the veil which thermal motion at normal temperature spreads over the inner world of atoms and electrons. . . . From each field of physics further questions push their way to the fore which are waiting to be solved by measurements at helium temperatures."

Liquid helium is often called a "quantum fluid." Upon one's first encounter with quantum physics, it may appear that its laws are important only for describing the properties of atoms, nuclei, electrons, and other microscopic particles. Liquid helium is a striking demonstration of the fact that quantum behavior may be manifested by macroscopic bodies as well.

It is now firmly established that helium is the only substance in nature which would not be solid at absolute zero (it solidifies only if subjected to strong external pressure). From the point of view of classical physics, the unique behavior of liquid helium is incomprehensible. Indeed, with decreasing temperature, the thermal oscillations of the particles become weaker and weaker, and the action of the intermolecular binding forces should lead to the material finally becoming a solid.

The behavior of liquid helium has nothing in common with this picture. Helium would remain a liquid even at absolute zero, even though at this point there would be no *thermal* motion at all.

During his very first experiment, Kamerlingh-Onnes was amazed at the very low density of liquid helium. It turned out to be eight times lighter than water! This very low density indicates that the light and inert helium atoms are also widely separated. Such a liquid is much harder to solidify than usual ones.

That is why helium remains a liquid down to extremely low temperatures. Furthermore, near absolute zero the laws of quantum physics prevent it from becoming a solid. According to these laws, the usual concept of atoms completely at rest at absolute zero is incorrect. We shall talk more about this subject, as well as about another remarkable property of liquid helium, superfluidity, in Chapter 10.

Having attained record low temperatures and having obtained liquid helium, Kamerlingh-Onnes now changed the direction of his research. He saw that the region of temperatures close to absolute zero contained an entire world of unique physical phenomena, and he undertook a systematic study of the properties of matter at low temperatures.

THE DISCOVERY OF SUPERCONDUCTIVITY

At the turn of the century, solid-state physics was in its infancy. One of the principal characteristics of a metal is its electrical resistivity, and physicists were very much interested in the dependency of this quantity on temperature. Kamerlingh-Onnes undertook such a study at liquid helium temperatures. He chose mercury as his sample, because it could be obtained in a sufficiently pure form.

In 1911, Kamerlingh-Onnes discovered the phenomenon of superconductivity. This phenomenon is actively studied to the present day and is one of the major branches of condensed matter physics. Kamerlingh-Onnes discovered that at a temperature close to 4 K the electrical resistance of mercury abruptly vanishes (Fig. 1.1). In his own words,

> the experiment left no doubt that, as far as the accuracy of measurement went, the resistance disappeared. At the same time, however, something unexpected occurred. The disappearance did not take place gradually, but abruptly. From 1/500 the resistance at 4.2°K drops to a millionth part. At the lowest temperature, 1.5°K, it could be established that the resistance had become less than a thousand-millionth part of that at normal temperature.
>
> Thus the mercury at 4.2°K has entered a new state, which, owing to its particular electrical properties, can be called the state of superconductivity.

The phenomenon of superconductivity is manifested in the electrical resistance vanishing at a finite temperature (called the critical temperature and denoted T_c). Interestingly, adding impurities to mercury did not destroy superconductivity, so that the original reasons for which Kamerlingh-Onnes chose mercury turned out to be unimportant.

One may wonder how an experiment can show that the resistance is identically zero. All measuring devices have a limit to their sensitivity, experimental uncertainties are always present, and so experimentally it is only possible to establish an upper limit on the observed resistance. This limit, however, turns out to be exceedingly small. The latest data show that the resistivity of a superconductor is below 10^{-27} Ω-cm (for comparison, recall that the resistivity of copper, which is an excellent conductor, is 10^{-9} Ω-cm), so there is no doubt that we

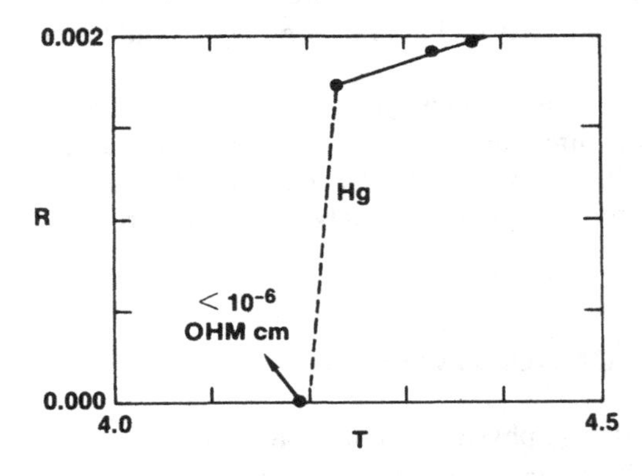

Figure 1.1. Discovery of superconductivity (H. Kamerlingh-Onnes, 1911): resistance of mercury versus temperature.

are dealing with ideal conductivity, that is, total vanishing of electrical resistance.

If one takes a ring made of a metal in the superconducting state and induces in it an electric current, the absence of resistance will result in the current not being damped and forever circulating around the ring. For instance, an experiment by Collins (in 1959) (see, e.g., Tinkham, 1975) showed that even after two and a half years there was no change in the current circulating around such a ring.

Kamerlingh-Onnes's discovery was followed by a large amount of experimental studies. New superconducting materials were found and their physical properties studied.

The absence of any resistance is a fundamental characteristic of superconductors but not the only one. They possess anomalous magnetic, thermal, and other properties, so it is actually more precise to talk not just about superconductivity, but about a peculiar state of matter observed at low temperatures.

SUPERCONDUCTING MATERIALS

Table 1.1 lists the elements which become superconducting at low temperatures, together with their critical temperatures. As far as is known today, the pure element with the highest transition temperature is niobium (T_c = 9.26 K) and the one with the lowest is tungsten (T_c = 0.012 K).

The critical temperature depends not only on the chemical composition, but on the crystal structure as well. Many substances can exist in several distinct crystalline forms which differ in their properties. For instance, gray tin is a semiconductor, while white tin is a metal, which, in addition, becomes a superconductor at a temperature of 3.72 K. There exist two different crystal structures of lanthanum (α-La and β-La), each with its own critical temperature. (4.9 K for α-La and 6.06 K for β-La). Beryllium is interesting in that it becomes a superconductor only if it is prepared as a thin film. Some elements turn into superconductors only at high pressures (for example, Ba becomes superconducting at a pressure of ~150 kbar, with $T_c \approx 5$ K).

It follows that superconductivity is not a property of isolated atoms, but is a collective effect determined by the structure of the whole sample.

It might be expected that good conductors such as copper, gold, and silver, which already have low resistance under normal conditions, would also be the first to become superconducting. But in fact something completely different is observed: in these metals there is no superconductivity at all. We will come back to this curious fact later.

The majority of superconductors are not pure elements, but alloys and compounds. Today over 6000 superconducting materials are known, and this number is constantly growing. There are superconducting alloys (such as CuS

Table 1.1. Values of T_c and H_c for the Elements[a]

Element	T_c (K)	H_c (G)[b]
Al	1.196	99
Cd	0.56	30
Ga	1.091	51
Hf	0.09	—
α-Hg	4.15	411
β-Hg	3.95	339
In	3.40	293
Ir	0.14	19
α-La	4.9	798
β-La	6.06	1096
Mo	0.92	98
Nb	9.26	1980
Os	0.655	65
Pa	1.4	—
Pb	7.19	803
Re	1.698	198
Ru	0.49	66
Sn	3.72	305
Ta	4.48	830
Tc	7.77	1410
Th	1.368	162
Ti	0.39	100
Tl	2.39	171
α-U	0.68	—
γ-U	1.80	—
V	5.30	1020
W	0.012	1
Zn	0.875	53
Zr	0.65	47

and Au_2Bi) and polymers $(SN)_x$ whose components by themselves do not have this property. This emphasizes once more that superconductivity is a collective phenomenon. Just as for pure elements, the critical temperature of alloys depends on their crystal structure. For example, the alloy Bi_2Pd has two very different transition temperatures (1.70 K and 4.25 K), depending on the lattice structure. If one or more of the components in an alloy is a superconducting element, the critical temperature of the alloy is different from that of its components, and is often higher.

The highest transition temperatures are in fact observed in alloys and compounds. For many years, the record holder was a niobium–tin alloy, with a critical temperature of 18.1 K. In 1973, it was discovered that films made out of the compound Nb_3Ge become superconducting at $T_c = 22.3$ K. In Table 1.2, the critical temperatures of some compounds are listed.

Figure 1.2 shows how the known transition temperatures increased with time. In the 75 years following the discovery of superconductivity, T_c increased by approximately 18 K [from T_c(Hg) $\approx$ 4 K to $T_c(Nb_3Ge) \approx$ 22 K]. The same figure illustrates the important role that niobium has played in the search for materials with higher T_c.

In 1986, the 75th anniversary of the discovery of superconductivity was marked by the discovery of a new class of superconducting materials, namely, copper oxides. A. Bednorz and K. A. Müller (IBM, Zurich) discovered superconductivity in the La–Ba–Cu–O system. One such compound remains superconducting up to 40 K.

The discovery of superconducting cuprates was followed by research growth at a rate unprecedented in the history of science. In early 1987, a group headed by C. W. Chu and M. K. Wu announced that a compound in the Y–Ba–Cu–O system becomes superconducting at a temperature higher than that of liquid nitrogen: its T_c is close to 100 K. As of this writing, the record is held by a compound in the Tl–Ca–Ba–Cu–O system, with $T_c \approx$ 125 K (Hermann and Sheng, 1988). The active search for new superconducting materials continues, and it is quite possible that soon even higher T_c's will be found.

Thanks to these recent advances, superconductivity is no longer just a low-temperature phenomenon. The discovery of high-temperature superconductivity has opened a new chapter in condensed matter physics. We will discuss the properties of the new superconducting oxides in Chapter 13. Now let us go on to describe another basic property of superconductors—the Meissner effect.

Table 1.2. Critical Temperatures of Selected Compounds

Compound	T_c (K)	Compound	T_c (K)
Nb_3Sn	18.05	$Pb_{0.7}Bi_{0.3}$	8.45
Nb_3Ge	22.3	V_3Si	1.7.1
NbN	16	$(SN)_x$	0.26
NbO	1.2	$(BEDT)_2Cu(NCS)_2$	10
$BaPb_{0.75}Bi_{0.25}O_3$	11	$La_{1.8}Sr_{0.2}CuO_4$	38
UBe_{13}	0.75	$Bi_2CaSr_2Cu_2O_{8+x}$	90

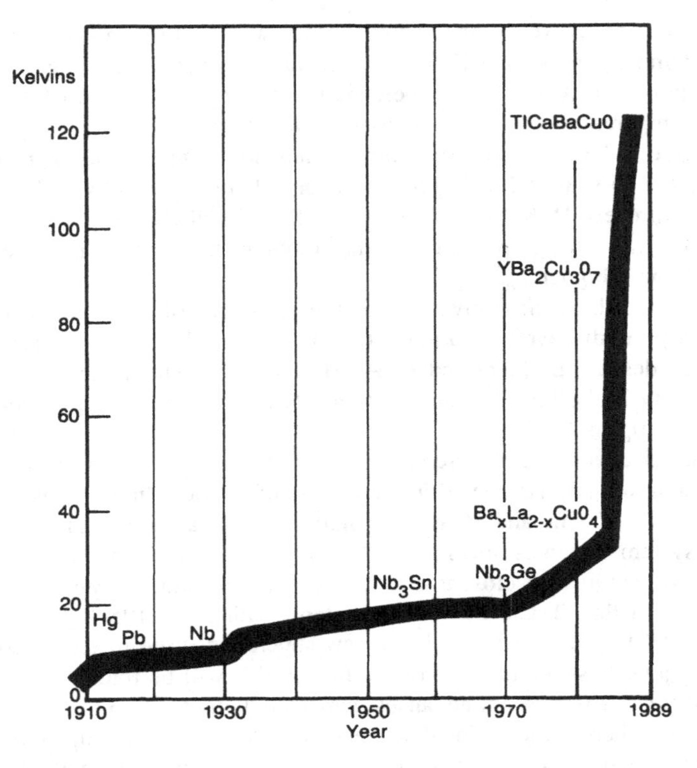

Figure 1.2. Increase in maximum value of T_c with time.

SUPERCONDUCTORS IN A MAGNETIC FIELD

In 1933, Meissner and Ochsenfeld discovered one of the most fundamental properties of superconductors. They found that magnetic fields do not penetrate into a superconducting sample. At temperatures above T_c, just as in any normal metal in an external field, there will be a finite magnetic field inside the sample. Let us start decreasing the temperature without removing the external field. We will find that at the moment the superconducting transition occurs, the magnetic field will be expelled from the sample (Fig. 1.3), and we will have $\mathbf{B} = 0$ (where $\mathbf{B}$ is the magnetic induction, defined as the average magnetic field in matter).

All metals other than ferromagnetics have zero magnetic induction in the absence of an external field. This is due to the fact that the magnetic fields created by the elementary currents in a metal are oriented chaotically and cancel.

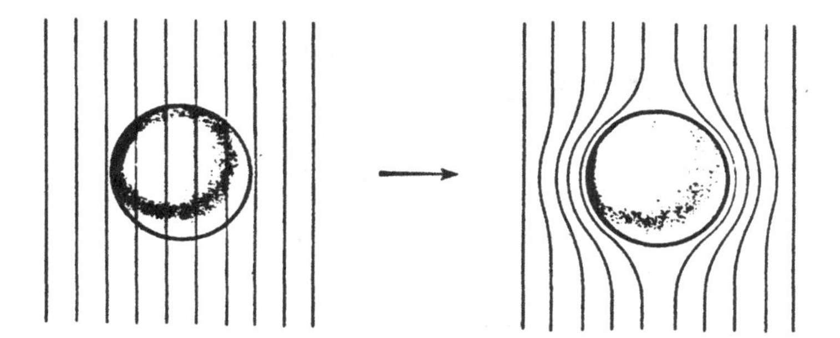

Figure 1.3. Meissner effect.

When an external field **H** is applied, there appears a finite induction **B** given by $\mathbf{B} = \mu\mathbf{H}$.

The coefficient μ is called the magnetic permeability. When $\mu > 1$ (paramagnetics), the applied field is enhanced. In diamagnetics ($\mu < 1$), the applied field is weakened and $\mathbf{B} < \mathbf{H}$. In superconductors, $\mathbf{B} = 0$, corresponding to zero magnetic permeability. This effect is called ideal diamagnetism.

How does the ideal diamagnetism in superconductors arise? It turns out that if a superconducting sample is placed in an external field, an electric current will be set up in the surface layer in such a way that its magnetic field will oppose the applied field. As a result, there will be zero magnetic induction deep inside the sample.

As is known, the external magnetic field which brings about the surface current does not do any work on the charged particles. In normal metals, an electric field is needed to do the work and sustain the current flow; there is none present here. Consequently, the ideal diamagnetism of superconductors implies that the surface current can flow without any resistance. Any such resistance would lead to thermal losses and, in the absence of an electric field, to quick damping of the current.

Thus, the Meissner effect and the phenomenon of superconductivity, that is,

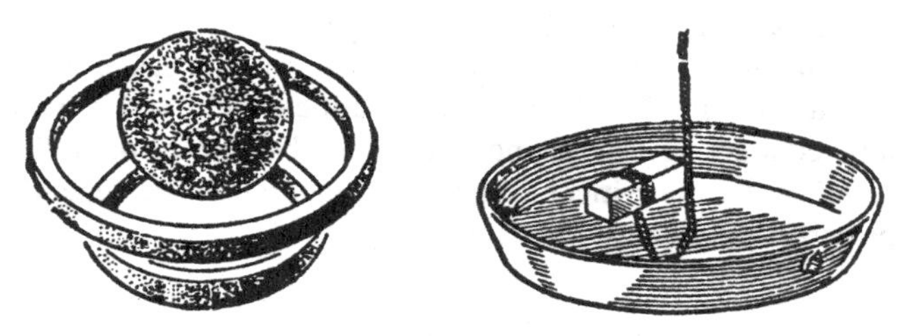

Figure 1.4. Levitation.

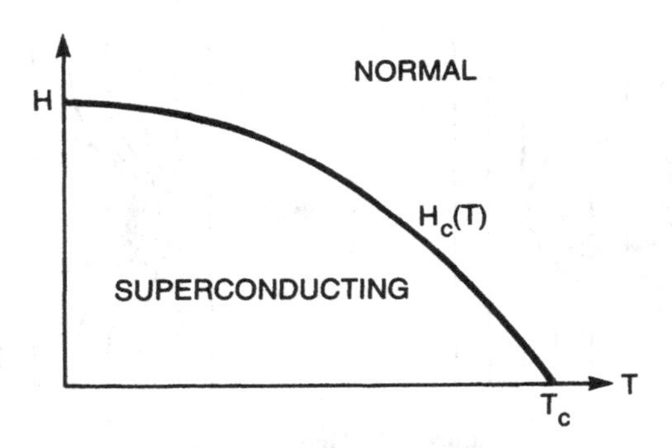

Figure 1.5. Critical field versus temperature.

the complete absence of resistance, are intimately related. They both have the same origin, established by the theory of superconductivity.

Let us also mention the following remarkable experiment. Take a metal ring with a current circulating around it, and place a superconducting sphere above this ring (Fig. 1.4 left). When the temperature is lowered below T_c, the sphere will become suspended above the ring ("levitation"). Alternatively, one can suspend a permanent magnet above a superconducting ring (Fig. 1.4 right).

As stated above, magnetic fields do not penetrate into a superconducting sample. This does not apply to the surface layer, where there will be a finite magnetic field. This layer contains the persistent currents which screen out the external field. The field penetration depth δ, that is, the thickness of the layer in which the magnetic field significantly differs from zero, is one of the main characteristics of a superconductor. When the decay of a field inside the sample is exponential, δ is defined as the distance it takes for the field to decrease by a factor of e. Usually, the penetration depth is a few hundred angstroms. The field thus penetrates into a superconductor to a depth of a few hundred interatomic distances.

If we start increasing the strength of the external field, then at some value (called the critical field, H_c) superconductivity is destroyed and the sample goes over into the normal state. The critical field depends on temperature. The closer we are to the critical temperature, the lower is H_c and the easier it is to destroy superconductivity. Figure 1.5 shows that the critical field varies with temperature. The superconducting state is most stable at absolute zero; at $T = 0$, H_c is maximum. At $T = T_c$, the critical field is zero, naturally. The dependence of H_c on T is approximately given by the formula

$$H_c(T) = H_{c0}\left[1 - \left(\frac{T}{T_c}\right)^2\right] \tag{1.1}$$

where H_{c0} is the value of the critical field at zero temperature. Table 1.1 lists the critical fields H_{c0} for some materials.

It is thus possible to convert a metal from the superconducting into the normal state not by raising its temperature but by subjecting it to a magnetic field.

THERMAL PROPERTIES

We have already stressed a number of times that superconductivity is not only the absence of electrical resistance, even though this property may be the most spectacular. In fact, we are dealing with a peculiar state of matter which displays many anomalous physical properties. As another illustration of this remark, we now consider the electronic heat capacity of superconductors.

The heat capacity of a metal is made up of contributions due to the electrons and to the crystal lattice. At low temperatures, it is mostly the electrons that contribute. With decreasing temperature, the electronic heat capacity decreases linearly, $c_{\text{en}} \sim T$, whereas the lattice heat capacity c_{lat} decreases much faster ($\sim T^3$) and does not play an important role for $T \rightarrow 0$.

The picture is substantially different in a superconductor. The electronic heat capacity c_{es} does decrease with temperature and vanish in the limit $T \rightarrow 0$; however, it does so not linearly, but exponentially (Fig. 1.6). This behavior can be written

$$\frac{C_{\text{es}}}{\alpha T_c} = ae^{-b(T_c/T)} \tag{1.2}$$

where a and b are constants independent of the temperature.

At $T \rightarrow 0$, superconducting substances are very sensitive to heat transfer processes. For example, even rather small heat losses can lead to strong cooling.

At the critical temperature, there is a transition into the normal state and the

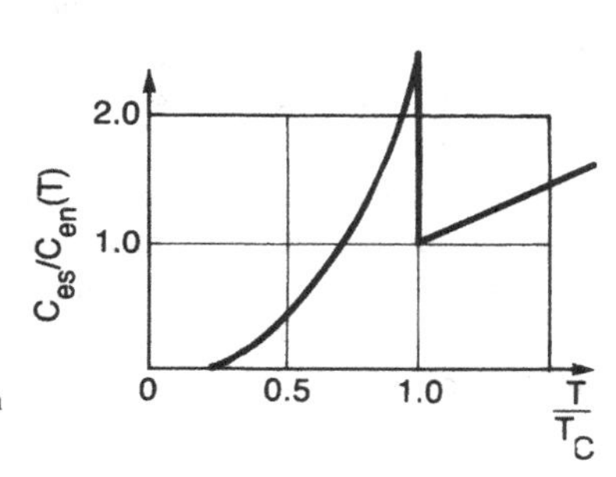

Figure 1.6. The electronic part of the heat capacity in the normal and superconducting states.

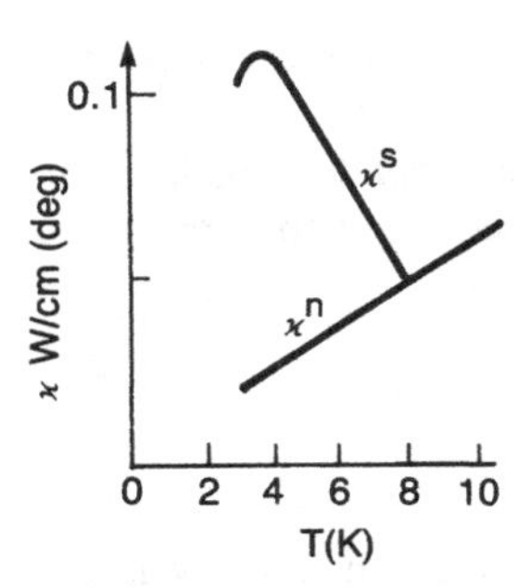

Figure 1.7. Thermal conductivity of the Pb + 10% Bi alloy.

electronic heat capacity begins to vary linearly. This transition is not smooth. At $T = T_c$, there is a jump in the heat capacity (Fig. 1.6), and $c_{es}(T_c)$ turns out to exceed $c_{en}(T_c)$. There is some variation in the magnitude of this jump from one superconductor to another (for instance, in Al, $c_{en}/c_{en} \approx 2.4$; in Pb, this jump is equal to 3.4), but it exists in all of them.

Heat transfer in superconductors is also characterized by some peculiar behavior. The temperature dependence of thermal conductivity in both normal and superconducting substances has been extensively studied experimentally. The thermal conductivity of a normal metallic alloy is known to decrease with decreasing temperature. Following the superconducting transition, there is an abrupt change: for $T < T_c$, the thermal conductivity rises sharply, then passes through a maximum, and only after that begins to drop (see Fig. 1.7).

We will analyze the thermal properties of superconductors in some detail in Chapter 5. Here we would like to note once again that these experimental features reflect the peculiar character of the superconducting state, whose physical properties are dramatically different from those of a normal metal.

THE MYSTERY OF SUPERCONDUCTIVITY

The discovery of superconductivity aroused an enormous interest. Many physicists tried their hand at explaining the disappearance of electrical resistance and other features of the superconducting state. Experimental information was accumulating, a necessary condition for resolving the puzzle.

A number of important phenomenological theories were created which allowed a number of phenomena to be described and predicted.

In 1934, Gorter and Casimir proposed the "two-fluid" model. This model assumes that the electron liquid in a superconductor can be imagined as a combination of two liquids: "normal" and "superconducting." The properties of the "normal" component are identical to those of the electron system in a normal metal, and the "superconducting" component is responsible for the anomalous properties. With decreasing temperature, the "superconducting" fraction grows, while the density of the "normal" component decreases and finally vanishes at zero temperature. At the phase-transition point, $T = T_c$, it is the "superconducting" part that vanishes. This latter component is responsible for the persistent currents, whereas the electronic heat capacity and thermal conductivity are governed by the "normal" fraction.

The two-fluid model proved to be a useful concept for analyzing the thermal properties of superconductors (see Chapter 5). Its physical meaning became clear after the formulation of the microscopic theory of superconductivity; we will come back to this question in Chapter 2. Let us point out right away that we enclose the words normal and superconducting in quotation marks here on purpose. The fact of the matter is that one may not interpret the existence of two such components as a real division of the electron system into two parts. Each separate electron cannot be either normal or superconducting. Superconductivity is a collective phenomenon, and it is more precise to talk not of two parts of the electron liquids, but of two types of motion the latter is capable of.

An important role was played by the London equation (1935), which provided a description of the anomalous diamagnetism of superconductors in a weak external field. Further generalization to the case of strong fields was achieved in the phenomenological Ginzburg–Landau theory (1950), based upon the Landau theory of second-order phase transitions. The Ginzburg–Landau theory played an important role in understanding the physics of the superconducting state. The equations derived from this theory (see Chapter 5) are highly nontrivial. Their validity was proven later on the basis of the microscopic theory.

The Ginzburg–Landau theory has made it possible to study the behavior of superconductors in strong magnetic fields. At first, it may seem that superconducting current is ideally suited for the production of strong magnetic fields. Indeed, when a normal conductor is used, a large part of the power goes into thermal losses. In order to obtain very strong fields, it is necessary to keep increasing the current, and the metal may simply melt as a result. There is no such danger in superconductors due to the absence of electrical resistance. It would seem, therefore, that we should be able to obtain magnetic fields of any desired magnitude by simply turning up the superconducting current. Unfortunately, as soon as the field becomes equal to H_c, superconductivity is destroyed. Nevertheless, a class of substances has been found that remain superconducting in strong magnetic fields and at intense currents. This subject, which is very important for practical applications, will be discussed in Chapter 5.

We shall not pause to discuss the phenomenological theories here. First, their important features will be described later on, on the basis of the modern theory of superconductivity. Second, this book is not intended to be a history of superconductivity, although we would like to hope that such a history will be written soon.

The phenomenological theories played (and continue to play) an important role, but the main problem of explaining the nature of superconductivity remained unsolved.

Many outstanding scientists grappled with this problem. They included Einstein, Heisenberg, Landau and Bloch, Frenkel, and others. Einstein, for example, pointed out an analogy between superconductivity and ferromagnetism. In ferromagnetics, the collective interaction between electrons leads to spontaneous magnetization which is stable with respect to thermal motion. Einstein attempted to construct a theory in which the collective interaction similarly leads to a formation of certain entities (Einstein called them "clouds") moving without friction. Landau and Bloch looked for current-carrying states which would be more energetically favorable. Frenkel studied the magnetic interaction between electrons. In 1947, Heisenberg published a series of papers in which he attempted to explain superconductivity by analyzing the Coulomb interaction in a many-electron system such as a metal. He considered a free-electron gas. At about the same time, Born and Cheng looked at a similar mechanism related to the Coulomb interaction; they also took into account the presence of a periodic crystal lattice.

All these theories now belong to the history of science: they were unsuccessful. We mentioned them here to show how difficult the problem was even for the best minds in physics.

It was only in 1957 that a microscopic theory of superconductivity was formulated. It took almost half a century to explain the phenomenon discovered in 1911. It would be difficult to think of another such stubborn mystery in the history of modern science.

THE ISOTOPE EFFECT

In this section we will describe an effect which played what was probably the decisive role in showing the way to the correct theory of superconductivity. All the other experimental facts presented above, in spite of their uniqueness, gave insufficient material for understanding the nature of superconductivity. Both experimentalists and theoreticians did their best to find out in which direction they should search for the explanation of this mysterious phenomenon.

The isotope effect, which had been studied theoretically by Fröhlich and discovered in 1950 by Maxwell and Reynolds, revealed this direction.

A study of different superconducting isotopes of mercury established a relationship between the critical temperature and the isotope mass; it turned out that as the mass number M was varied from 199.5 to 203.4, T_c changed from 4.185 to

4.140 K. With sufficient precision, it was established that for a given element the following relationship holds:

$$T_c M^{1/2} = \text{const.} \tag{1.3}$$

What does this result mean? The isotope mass is a characteristic of the crystal lattice and can affect its properties. For example, the frequency of lattice vibrations is related to the ion mass in the following way: $\Omega \sim M^{-1/2}$. Superconductivity, which is a property of the electron system, is shown by the isotope effect to be related to the state of the crystal lattice. It follows that superconductivity is due to the interaction between the electrons and the lattice. Interestingly, this very interaction is also responsible for the electrical resistance. Under certain conditions, it leads to the disappearance of the latter, that is, to superconductivity.

Apparently, the fact that electron–phonon scattering is one of the principal mechanisms of resistance was for many years a stumbling block, making it extremely hard to imagine that the same interaction can lead to a vanishing of resistance.

The isotope effect was discovered in mercury, as we have already said. Curiously, this substance served for finding both the long-mysterious phenomenon of superconductivity (1911) and the isotope effect which helped to crack the puzzle. The theory of superconductivity was developed soon after the discovery of this effect.

2

THE NATURE OF SUPERCONDUCTIVITY

MICROSCOPIC THEORY

A systematic theory of superconductivity, which explained the nature of this phenomenon, was formulated in 1957 by Bardeen, Cooper, and Schrieffer (BCS theory). The mystery of superconductivity was solved, and the formulation of the theory brought further progress in the field. In this section and the next, we will discuss the basic principles of this theory.

The discovery of the isotope effect clearly indicated that, in an explanation of the phenomenon of superconductivity, the interaction between electrons and the crystal lattice must be taken into account. It turns out that this interaction leads to attraction between the electrons in a superconductor. At first glance, this seems strange, since we are used to seeing electrons, having the same charge, repel each other. In fact, there is no contradiction with the laws of physics here. Of course, two electrons in a vacuum will repel each other. We are not considering, however, the problem of two isolated electrons. The electrons are located in a medium, that is, inside a crystal, and the presence of a medium can change the sign of the interaction.

What sort of mechanism, then, is involved in the interelectron attraction? It turns out that the interaction between the electrons and the lattice leads to a certain effective interaction between the electrons. The nature of this interaction is as follows. An electron moving in a metal deforms, or polarizes, the crystal lattice by means of electric forces. The displacement of the ions in the lattice caused in this way affects the state of the other electron, since the latter now finds itself in the field of the polarized lattice with a somewhat altered structure. This results in an effective attraction between the electrons (Fig. 2.1).

The appearance of the attractive force can be visualized as follows. As a result of the deformation of the lattice, an electron is surrounded by a "cloud" of

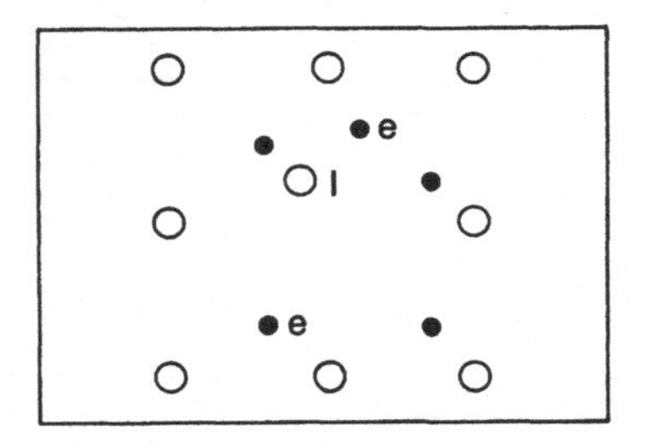

Figure 2.1. Additional effective interaction between the electrons. The displacement of the ion *I* caused by its interaction with the electron *e'* affects the state of the other electron.

positive charge which is attracted to the electron. The magnitude of this positive charge can exceed the electron charge. Then this electron, together with the surrounding "cloud," represents a positively charged system, which will be attracted to another electron.

At high temperatures, sufficiently intense thermal motion pushes particles apart from each other and washes away the ion "coat," which effectively reduces the attractive forces. At low temperatures, however, these forces play a very important role.

The crystal lattice is thus the medium which makes the dielectric permittivity in a superconductor negative.

The appearance of the additional interelectron attraction can be described in quantum terms as well. Let us consider a metal at $T = 0$ K. Its crystal lattice is not at absolute rest, but executes so-called "zero-point" vibrations, which exist due to the quantum-mechanical uncertainty principle and correspond to the ground state of a harmonic oscillator. An electron moving in the crystal disrupts the state of these zero-point vibrations and excites the lattice. When the lattice returns into its ground state, it radiates energy that is absorbed by another electron. The excited state of the crystal lattice is described in terms of sound quanta—phonons. Therefore, the process described above can be visualized as the emission of a phonon by an electron moving through the lattice and the subsequent absorption of the phonon by another electron. This exchange of phonons is what leads, in the quantum picture, to the additional attraction between electrons.

At low temperatures, this attraction in a number of materials (these are superconductors) prevails over the Coulomb electron repulsion. Here the electron system turns into a bound unit, and finite energy must be expended in order to excite it. The energy spectrum of the electron system (we stress that we are referring not to the energy of individual electrons, but to the energy of the entire system of bound electrons, treated as a single unit) in this case will not be continuous. The excited state of the electron system is separated from the ground state by an energy interval or, as it is referred to, by the energy gap Δ.

Discrete energy spectra are common in quantum physics. They arise whenever we are talking about energy levels of electrons in an isolated atom or molecule. In the case of a superconductor, however, we are dealing with discreteness in a *macroscopic* electron system.

It should be noted that in quantum theory of the solid state, there already exists the concept of a forbidden band. For example, in an ordinary semiconductor at $T = 0$ K, there is a gap between the filled valence band and the empty conduction band. This sort of situation corresponds in the band picture to the absence of free carriers. A superconductor does have carriers, though, which can carry the superconducting current. However, these carriers are in a bound state, and finite energy must be expended in order to excite this state (for example, by means of incident electromagnetic radiation).

It is known that the electrical resistance of a metal is due to the interaction of the moving electron system (that is, the electric current) with the vibrations of the crystal lattice or with impurities. It is clear, however, that if there is a gap in the energy spectrum, quantum transitions in the electron fluid will not always be possible. The electron system will not be excited when it is moving slowly. This implies movement without friction, that is, the absence of electrical resistance.

If the current exceeds a certain critical value (the "critical current"), then superconductivity disappears.

The interaction between electrons and the crystal lattice is one of the basic mechanisms of electrical resistance in an ordinary metal. It is remarkable that it is precisely this interaction that, under certain conditions, leads to an absence of resistance, that is, to superconductivity. This is why the seemingly strange behavior of such metals as gold, silver, and copper in fact is not surprising at all. These metals are excellent conductors under ordinary circumstances but do not display superconductivity. But good conductivity, after all, indicates that the electrical resistance is low and, therefore, that the interaction between electrons and the lattice is very weak. Near absolute zero, such a weak interaction does not create sufficient interelectron attraction to overcome the Coulomb repulsion. Therefore, there is no transition into the superconducting state.

At absolute zero, electrons in a metal occupy energy states up to the Fermi level, E_F. The equation $E(\mathbf{p}) = E_F$ defines the so-called Fermi surface in $\mathbf{p}$-space. In the simplest case, when $E = p^2/2m^*$ (m^* is the electron "effective" mass), the Fermi surface is a sphere.

The presence of attraction leads to a restructuring of the Fermi surface because of the appearance of the energy gap. Whereas in a normal metal the electron energy, referred to the Fermi surface, is equal to $\xi = p^2/2m^* - E_F$, and thus can be arbitrarily small, in superconductors it is described by the expression

$$\epsilon = (\xi^2 + \Delta^2)^{1/2} \tag{2.1}$$

From this formula, it is clear that the minimum energy, corresponding to $\xi = 0$, is nonvanishing and is equal to Δ. The electrons in a layer of thickness $\Delta(0)$ near the Fermi surface form bound pairs.

The magnitude of the energy gap $\Delta(0)$ (we are still considering systems at absolute zero of temperature) depends on the strength of the electron–phonon coupling λ; this dependence has the following peculiar form:

$$\Delta = 2\hbar\,\tilde{\Omega}\,e^{-1/\lambda} \tag{2.2}$$

where $\tilde{\Omega} \sim \Omega_D$, the Debye frequency. The gap thus increases following a curious exponential law. Looking at this simple formula, one recognizes what was one of the major stumbling blocks on the road to the theory of superconductivity. It had to be overcome (and this was done in the BCS theory) even after it became clear that superconductivity was caused by the electron–lattice interaction. The origin of the difficulty is that the function $y = e^{-1/x}$ has an interesting property: it cannot be expanded in a Taylor series for small x (the reader can easily verify this). On the other hand, as is well known, one of the main tools of quantum mechanics is perturbation theory. In this method, such quantities as the wave function and energy are obtained as expansions in powers of some small parameter (perturbation). It would seem that perturbation theory is perfectly suited for the case when the electron–phonon interaction is weak. However, attempts along these lines failed; the BCS theory as well as its subsequent versions use other approaches. Today it is clear that perturbation theory would be unable to predict a dependence of the form represented by Eq. (2.2) even in the case of weak coupling (i.e., small λ) because the function in Eq. (2.2) cannot be represented as a power series.†

Thus, the effect of superconductivity arises as a result of interelectron attraction. This attraction possesses certain peculiar features. Let us suppose that we are interested in the behavior of some electron in a superconductor. It turns out that this electron will be attracted unequally to all the other electrons. It will "choose" one specific electron (this "chosen" electron will have opposite momentum and spin), and the interaction between this pair will be the strongest. The electron system in a superconductor can be described as consisting of bound pairs of such electrons, and an excitation of the electron system can be described as a

†We are not going to describe in detail the theoretical approach developed in the BCS theory, because this is outside the scope of this book. The interested reader can learn about it from specialized sources (see the list of recommended literature). We should mention that after the appearance of the BCS theory, other important methods were developed by Gor'kov (see Abrikosov *et al.*, 1963) and by Bogoliubov. Gor'kov's method is very useful for analysis of spatially inhomogeneous systems (behavior in external fields, properties of alloys, etc.), while Bogoliubov's approach is convenient, for example, for calculating kinetic coefficients.

breakup of such a pair. The quantity $2\Delta(0)$, where $\Delta(0)$ is the energy gap defined by Eq. (2.2), is the pair binding energy.

The state of electrons in a metal is changing constantly, and, as a result, the sets of pairs are also changing continuously. One could say that it is as if the electrons in a superconductor have come together for a huge gala ball and are performing a dance in which they are constantly changing partners.

Electron pairs are often called Cooper pairs, after Leon Cooper, who first demonstrated that weak attraction between electrons in a metal leads to their bound state.

Let us pause to appreciate the significance of Cooper's theorem. One might think that if there is an attractive force between particles, it will always lead to the formation of a bound state. Things are not quite that simple in quantum mechanics. In the language of quantum physics, the appearance of a bound state implies the existence of an energy level in the potential well which describes the attraction. It turns out that, in the usual three-dimensional case, such a level occurs only if the depth of the well exceeds a certain minimum value.

Cooper demonstrated, however, that the attraction between electrons in the presence of a Fermi sea formed by other electrons leads to the formation of a bound state no matter how weak the attraction. This statement can be understood as follows. The crux of the matter is that above we were talking about the three-dimensional case. If we were to consider a two-dimensional problem (i.e., solve the two-dimensional Schrödinger equation), the conclusion would be different. Namely, it would turn out that a bound state will arise for arbitrarily weak attraction.

Let us now come back to electrons in a metal. The electrons on the Fermi surface are executing essentially two-dimensional motion (in momentum space). Because of the Pauli principle, they cannot make transitions into states inside the Fermi surface. This reduced dimensionality makes the solution represented by Eq. (2.2) possible in the weak-coupling limit. In this connection, it is interesting to note that the corresponding two-dimensional Schrödinger equation indeed cannot be solved by perturbation theory, whereas an exact solution in fact gives an energy level of the type given by Eq. (2.2) (see Appendix A).

We should mention at this point that the high-temperature superconductors have a layered structure. The carrier motion there is quasi two-dimensional, which is favorable for pairing. It becomes possible even for carriers far from the Fermi level to bind, leading to a large value of the energy gap. In these materials (which we will discuss in some detail in Chapter 13), the ratio $\Delta(0)/E_F$, which indicates the fraction of the electrons that are paired, is much larger than in conventional superconductors.

The question of the size of an electron pair deserves particular attention. An estimate of this size can be made in the following manner. In the simplest case, the Fermi energy E_F is equal to $p_F^2/2m$. The superconducting state is characterized by the presence of a gap Δ near the Fermi surface. The momentum spread connected with this is determined by the relation $p_F\delta p/m \approx \Delta$, where $\delta p \approx m\Delta/p_F$. From the uncertainty relation $\delta p\delta r \approx \hbar$, one can estimate the spatial spread as $\delta r \approx \hbar/\delta p \approx \hbar v_F/\Delta$, where v_F is the Fermi velocity. It is convenient to introduce a numerical factor and then the coherence length (size of the Cooper pair) is described by the expression

$$\xi_0 = \frac{\hbar v_F}{\pi\Delta} \qquad (2.3)$$

The quantity ξ_o characterizes the scale of spatial correlation in a superconductor. Substituting typical values of v_F and Δ, we obtain $\xi_o \approx 10^{-4}$ cm.

For comparison, recall that the period of a crystal lattice is approximately 10^{-8} cm, so that the electrons forming the pair are separated by an immense distance of 10^4 lattice spacings. Nevertheless, they are the ones that are most strongly attracted to one another. If the state of one of the electrons making up the pair changes under the influence of some sort of force (for example, under the influence of a magnetic field), this change immediately has an effect on the behavior of the other electron. In superconductors, we encounter an example of long-range correlation, which is unique in the inorganic world.

We should note, however, that in the new high-temperature superconductors (Chapter 13), the size of pairs is relatively small, which is an important feature of these new materials.

FINITE TEMPERATURES. THE CRITICAL TEMPERATURE

In the preceding discussion, we were neglecting the thermal motion of the electrons. In essence, we were considering superconductivity at $T = 0$ K. Consider now the state of affairs at finite, nonzero, temperatures. Thermal chaotic motion excites the electron system and reduces interelectron attraction. We are thus faced with the problem of describing excited states, rather than the ground state, of a many-particle superconducting system.

Low-lying excited states of a many-body system can be described with the aid of the elementary excitation, or quasiparticle, method (see Appendix B).

At finite temperatures, in superconductors there appear quasiparticles which can change their energies by an arbitrary amount and thus behave just like ordinary electrons. They are described by the usual Fermi–Dirac distribution, but with a somewhat different expression for the energy, namely,

$$n = [e^{\epsilon/T} + 1]^{-1} \qquad (2.4)$$

where $\epsilon = [\xi + \Delta^2(T)]^{1/2}$; $\Delta(T)$ depends on temperature.

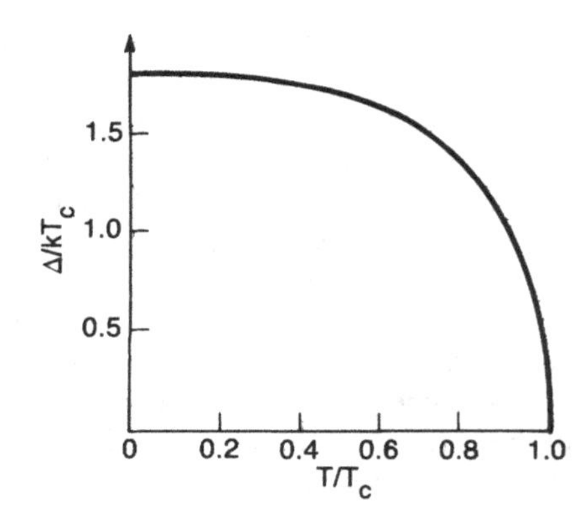

Figure 2.2. Energy gap versus temperature.

The excitation energy is seen to be described by a formula analogous to Eq. (2.2). The quantity $\Delta(T)$ plays the same role as the energy gap introduced at zero temperature (see above). We shall call it the temperature-dependent energy gap, although, strictly speaking, a gap in the energy spectrum can be defined only at $T = 0$ K.

The temperature dependence of $\Delta(T)$ is shown in Fig. 2.2. At $T = 0$ K, $\Delta(T)$ is equal to $\Delta(0)$. With increasing temperature, the gap decreases and finally vanishes at $T = T_c$. In the BCS theory, it is shown that

$$\Delta(T)|_{T \to T_c} = aT_c \sqrt{1 - \frac{T}{T_c}} \tag{2.5}$$

where, in the weak-coupling approximation, $a = a_{\mathrm{BCS}} = 3.06$.†

Equation (2.4) shows that, because of the term containing $\Delta(T)$, the number of excitations of a given energy is less than the corresponding number of electrons in a normal metal. As $T \to 0$, these excitations vanish altogether. At temperatures very close to absolute zero, $n \sim e^{-\epsilon/T}$, so that n is exponentially small. Conversely, for $T \geq T_c$, when $\Delta = 0$, the function given by Eq. (2.4) becomes the usual expression for the electrons in a normal metal.†† One can see directly from Eq. (2.4) that at temperatures below T_c, the presence of the gap $\Delta(T)$ leads to the number of quasiparticles (usually called simply "normal electrons") being less than the total number of electrons. We have thus arrived at the "two-fluid" model which describes the superconductor as containing "normal" electrons, that is, quasiparticles, and a "superconducting" component.

†We are using energy units ($k_B = 1$).

††We are describing the state of a normal metal in terms of electrons for $E > E_F$ and in terms of holes for $E < E_F$. Then the function $n = 2[\exp(|\xi|/T) + 1]^{-1}$ obtained from Eq. (2.4) with $\Delta = 0$ indeed corresponds to a normal metal.

The quasiparticle description implies that the separation into "normal" and "superconducting" components cannot be taken literally. The electrons in a superconductor are not divided up into "normal" and "superconducting." What is actually happening is that at finite temperatures the thermal motion reduces the superconducting ordering of electrons. The quasiparticle picture is used to describe this process. The quasiparticles correspond to, and determine the entropy of, the "normal liquid"-type motion in a superconductor.

The quasiparticle method thus justifies the two-fluid model, which provides an intuitive understanding of many experimental facts.

The energy gap $\Delta(T)$ decreases with increasing temperature. At some value of the latter, the gap vanishes and the material becomes normal. This value is the critical temperature. The critical temperature is described by the following fundamental expression:

$$T_c = 1.14\ \hbar\tilde{\Omega}e^{-1/\lambda} \tag{2.6}$$

where $\tilde{\Omega} \simeq \Omega_D$.

Let us make a few remarks about Eq. (2.6). First of all, we should stress that it is valid only in the weak coupling limit ($\lambda \ll 1$), just like Eq. (2.2) for the energy gap. If the coupling constant is not small, different expressions must be used (see Chapter 6).

In the BCS theory, the coupling constant λ is expressed as $\lambda = V\ \nu_F$; here V is the matrix element describing electron–lattice interaction, and ν_F is the density of states at the Fermi surface. For usual three-dimensional metallic systems, $\nu_F = (\pi^2/\hbar^3)^{-1}\ mp_F$, where m is the effective mass of the carriers and p_F is the Fermi momentum. The quantity λ is assumed to be independent of the lattice vibrational frequency, so that the critical temperature depends on the phonon frequency only via the preexponential factor in Eq. (2.6). A more detailed analysis of the electron–phonon interaction shows that this assumption would be valid only if the phonons had a simple acoustic character: $\Omega = uq$, where u is the sound velocity and q is the phonon momentum. In reality, as opposed to a model picture, the simple acoustic law holds in the long-wavelength part of the phonon spectrum. On the other hand, in the short-wavelength part, which contributes the most to the interelectron attraction, the dependence $\Omega(q)$ is not a simple linear one. As a result, λ turns out to depend on the phonon frequency (see Chapter 6). In fact, $\lambda \sim \Omega^{-2}$, where $\Omega = \langle\Omega^2\rangle^{1/2}$, with $\langle\ \rangle$ denoting the average value.

A small value of the coupling constant does not by itself mean that the critical temperature is also small. T_c is determined by two factors: the strength of the coupling (the constant λ) and the energy scale (the phonon energy $\hbar\Omega$). The combination of these two factors can lead to a large value of T_c even if λ is small. As we emphasized earlier, for larger λ values, one must employ different formulas.

In the BCS model, phonons do not appear explicitly. In their place, there is introduced some quantity which describes interelectron attraction and is assumed constant in the energy interval $0 < \epsilon < \Omega_D$. A more detailed analysis, to be described in Chapter 6, shows that the solution in Eq. (2.6) is indeed valid in the weak-coupling approximation; only the preexponential coefficient in Eq. (2.6) comes out different (in other words, the BCS solution has a preexponential accuracy).

From Eq. (2.6) it follows that $T_c \sim \tilde{\Omega}$. Since the vibrational frequency $\tilde{\Omega} \sim M^{-1/2}$, where M is the ion mass, we have $T_c M^{1/2}$ = const. This explains the isotope effect which played such a pivotal role in solving the mystery of superconductivity. A more detailed analysis has shown, however, that the isotope effect is quite a complicated phenomenon and that the exponent in the relation $T_c \sim M^{-\alpha}$ may be significantly different from 0.5 (see Chapter 6).

Thus far, we have concentrated on the role of the interelectron attraction. This attraction is characterized by the constant λ which enters into Eqs. (2.2) and (2.6). In fact a substance becomes superconducting if this attraction becomes sufficiently strong to overcome the Coulomb repulsion. The formula for T_c should be modified to reflect the presence of the repulsion.

Let V_c be the average Coulomb interaction. One is tempted simply to replace λ by the difference $\lambda - V_c$ in the formula for T_c and then to require $\lambda > V_c$. Actually, the situation is much more interesting. A remarkable feature of the theory of superconductivity is the so-called logarithmic reduction of the Coulomb repulsion. The fact of the matter is that the interelectron Coulomb repulsion is screened by the other electrons. As a result, the range of the repulsive forces becomes much smaller than the size of a pair. On the energy scale, the characteristic energy of the phonon-mediated attraction is the lattice vibrational energy $\hbar\tilde{\Omega} \sim \hbar\Omega_D$, whereas for the Coulomb repulsion the characteristic energy is on the order of the Fermi energy. A detailed analysis showed that in the theory of superconductivity the Coulomb repulsion is described not directly by V_c, but rather by the so-called pseudopotential μ^* given by

$$\mu^* = \frac{V_c}{1 + V_c \ln\left(\frac{\omega}{\Omega_D}\right)}; \qquad \omega \simeq E_F \tag{2.7}$$

The fact that $E_F \gg \Omega_D$ (usually, $E_F \approx$ 5–10 eV, while $\Omega_D \approx$ 0.01–0.05 eV) results in the inequality $\mu^* \ll V_c$. The critical temperature is determined by the difference $\lambda - \mu^*$, that is,

$$T_c = 1.14\,\hbar\tilde{\Omega}\,e^{-1/(\lambda-\mu^*)} \tag{2.8}$$

The quantity μ^* is usually small: $\mu^* \approx 0.1$. Naturally, the equation for the gap (Eq. 2.2) must be similarly modified.

Superconductivity is not a universal phenomenon. It shows up in materials in which attraction overcomes repulsion. At present, unfortunately, we still do not have an exact criterion that could establish what the structure and composition of a sample must be in order for attraction to dominate. The difficulty in formulating such a criterion is connected with the fact that the energy change during a superconducting transition is very small. The energy difference between the normal and superconducting states is equal to $H_c^2/8\pi$ per unit volume, where H_c is the critical field. (We have used the well-known formula $w = H^2/8\pi$ for the magnetic field energy density; recall that at $H = H_c$, the superconducting state disappears). This quantity corresponds to just 10^{-8} eV/atom, compared to $\sim$1 eV/atom for the interelectron Coulomb interaction. Therefore, it is first necessary to have an accurate description of the normal state.

Above, we gave the expressions for the gap at $T = 0$ K, $\Delta(0)$ (Eq. 2.2), and the critical temperature, T_c (Eq. 2.6). In the BCS theory, these quantities are proportional to each other (the law of corresponding states). The ratio of the two is a definite number, namely,

$$\frac{\Delta(0)}{T_c} = 1.76 \tag{2.9}$$

Formulas (2.5) and (2.9) show that the BCS theory, developed in the weak-coupling approximation, is characterized by universal relations, independent of the material, the specifics of its phonon spectrum, and so on. The same holds for many other formulas, making the model exceptionally elegant and complete. The universality is lost when the effects of strong coupling are analyzed.

3

ENERGY GAP

The presence of an effective interelectron attraction in superconductors leads to the appearance of a gap in the energy spectrum. As a result, the electronic system is unable to absorb arbitrarily small amounts of energy. The energy gap in superconductors can be directly observed [first observed by means of infrared spectroscopy (see Chapter 5) by Glover and Tinkham (Tinkham, 1975)]. Experiments are capable of not only demonstrating the existence of the gap (which in itself confirms the basic tenet of the theory of superconductivity) but also of measuring its magnitude. Below we shall consider several effects which provide direct evidence for the presence of the energy gap. These effects are observed both in superconductors and normal materials, but here we shall devote our attention to the features peculiar to superconductors.

TUNNELING

Consider the flow of electrons across a thin insulating layer about 10 Å thick which separates a normal film from a superconducting one. In the presence of a potential barrier, whose role in our case is played by the dielectric layer, there is a finite probability that electrons will tunnel across the barrier. This is an example of quantum-mechanical tunneling. It will give rise to a tunneling electric current. There is a nonzero tunneling probability if the electron ends up in an allowed state of energy lower than, or equal to, the energy of the initial state. Figure 3.1 shows a superconductor–normal metal tunneling junction.

The levels are filled up to the maximum—E_F—in the normal metal, and up to $E_F - \Delta$ in the superconductor (Δ is the energy gap). Under such circumstances, no tunneling current can appear. As Fig. 3.1 illustrates, because of the presence of the energy gap in the superconducting film, there are no states available for tunneling transitions.

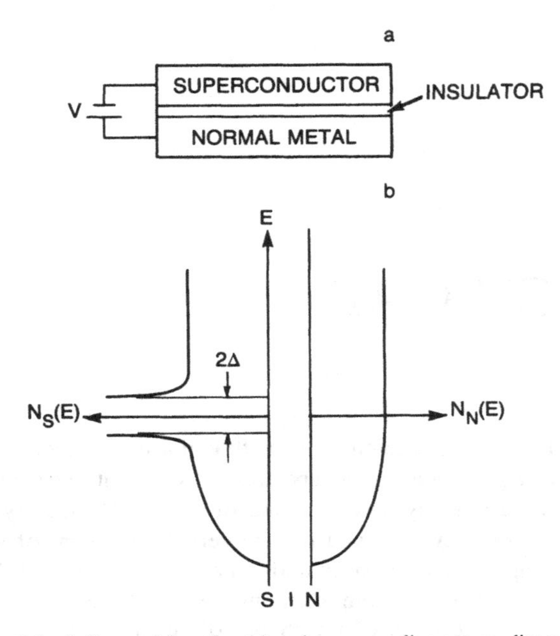

Figure 3.1. *S–N* tunnel junction (a) and corresponding energy diagram (b).

In order to make these transitions possible, it is necessary to place the system into an external electric field. The field will alter the distribution of the energy levels. Tunneling will become possible when the applied voltage reaches the value Δ/e. Figure 3.2 shows the observed current–voltage characteristic (the dotted line corresponds to a junction between two normal metals). It is seen that the tunneling current appears only at a finite voltage when the product eV becomes equal to the energy gap. At higher voltages, the curve approaches the dependence characteristic of a tunneling junction between two normal metals. The effect described here was first observed by Giaever in 1960.

The absence of a tunneling current at small voltages constitutes an experimental proof that there exists a gap in the energy spectrum of a superconductor. The magnitude of the gap can be measured with a voltmeter and corresponds to the voltage at which the tunneling current just appears. Table 3.1 lists the values of $2\Delta(0)/T_c$ for several metals, obtained in this way.

Let us consider the tunneling effect in more detail. Of course, tunneling can also take place when both films are in the normal state (an *N–I–N* junction; Fig. 3.3). We will be considering ordinary elastic processes in which the electron energy is conserved. Because of the Pauli principle, such transitions are permitted only if there is a vacant state of the required energy in one of the films. It is

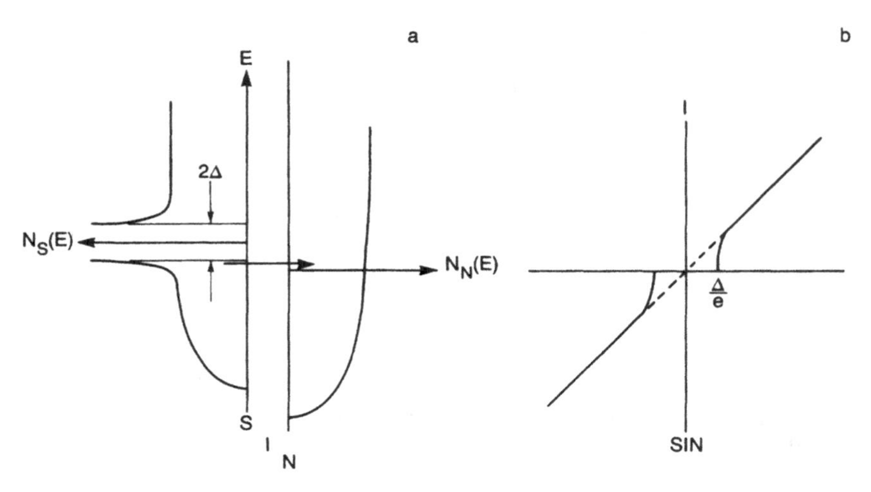

Figure 3.2. Energy diagram in the presence of applied field (a) and I–V characteristic (b) for S–I–N junction; the dashed line corresponds to N–I–N junction.

clear that without an externally applied field this is impossible, because when two metals are in thermal equilibrium, their Fermi levels coincide. In the presence of an applied potential V, there is a finite tunneling current, equal to

$$\mathbf{j} = \frac{4\pi e^2}{\hbar} \sum_{\bar{p},\bar{q}} |T_{\mathbf{pq}}|^2 \, \Phi_{\mathbf{p,q}} \, \delta(E_{\bar{q}} + V - E_{\bar{q}}) \tag{3.1}$$

Here the quantity $T_{\mathbf{pq}}$ (the so-called tunneling matrix element) describes the probability of tunneling, p and q are the electron momenta in the metals on the left-hand and the right-hand side, respectively, and

$$\Phi_{\mathbf{p,q}} = f_{\mathbf{q}} (1 - f_{\mathbf{p}}) - f_{\mathbf{p}} (1 - f_{\mathbf{q}}) \tag{3.2}$$

or

$$\Phi_{\mathbf{p,q}} = f_{\mathbf{q}} - f_{\mathbf{p}} \tag{3.3}$$

The first term in Eq. (3.2) describes electronic transitions from the left-hand film into the right-hand one. It is proportional to $f_{\mathbf{q}}$, the probability of there being an electron in the state **q**, and to $(1 - f_{\mathbf{p}})$, the probability of the state **p** being vacant. The second term in Eq. (3.2) describes transitions in the other direction, and thus the quantity $\Phi_{\mathbf{p,q}}$ corresponds to the net current.

Consider $T = 0$ K. Transitions mainly take place between states close to the Fermi level. We assume that in this region $T_{\mathbf{pq}}$ is a constant; that is, we replace it

Table 3.1. Selected Tunneling Data

Material	$2\Delta/T_c$
Al	3.5
Tl	3.6
Nb	3.9
Pb	4.3
Am–Mo	3.7
$Pb_{0.7}Bi_{0.3}$	4.86

by T. We also go from summation over energies to integration by introducing the density of states ν_N, that is, the number of quantum states per unit energy interval:

$$\nu_N = \frac{2dp_x dp_y dp_z}{(2\pi\hbar)^3\, dE} \tag{3.4}$$

In the isotropic case, $dp_x dp_y dp_z = 4\pi p^2 dp$, so that $\nu_N = (2\pi^2\hbar^3)^{-1}p^2(dp/dE) = (2\pi^2\hbar^3)^{-1}pm^*$, where m^* is the effective mass. For the density of states in Eq. (3.4), we can use its value at the Fermi surface; in the isotropic model

$$\nu_N^F = (\pi^2\hbar^3)^{-1}m^*p_F \tag{3.5}$$

Now, integrating over $E_\mathbf{p}$ and then over $E_\mathbf{q}$ (for small applied voltages, $eV \ll E_F$, we can limit ourselves to the linear term in the expression for the tunneling current), we obtain the following expression for the conductivity $\sigma_N = (\partial j/\partial V)|_{T=0\,\mathrm{K}}$ of the tunneling junction N_L–I–N_R:

$$\sigma_N = \frac{4\pi e^2}{\hbar}\, \nu_{NL}^F\, \nu_{NR}^F \tag{3.6}$$

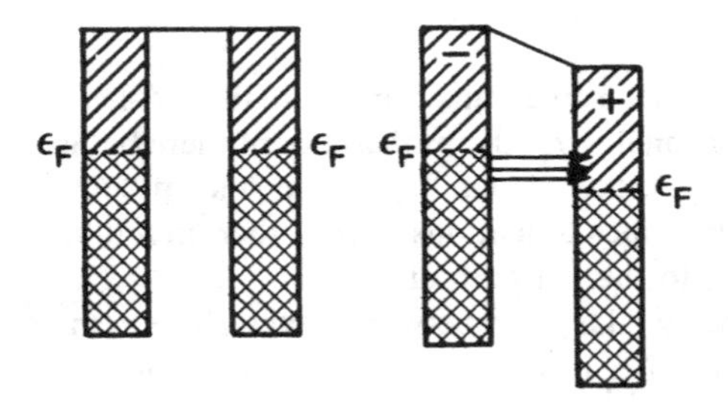

Figure 3.3. N–I–N junction: (a) absence of tunneling current due to the Pauli principle; (b) effect of applied voltage.

Here ν_{NL}^F and ν_{NR}^F are the densities of states at the Fermi level for the left-hand and right-hand films, respectively. It is seen that the tunneling current j is proportional to the applied voltage V, with conductivity given by Eq. (3.6).

Now consider the case when one of the films is a superconductor (an N–I–S junction). The calculation of the conductivity is largely analogous to that given above, but there is one crucial difference. The density of states ν_S in a superconductor is significantly different from ν_N; this is due to the form of the excitation spectrum in superconductors [Eq. (2.1)]. The density of states ν_S is equal to $(\partial n/\partial \epsilon)$, where $n = 2dp_x dp_y dp_z (2\pi\hbar)^{-3}$ is the number of quantum states (per unit volume) and ϵ is the excitation energy in a superconductor. As a result, we find $\nu_S = \nu_N \epsilon/\xi$, or $\nu_S = \nu_n \epsilon (\epsilon^2 - \Delta^2)^{-1/2}$. It can be seen that the density of states in a superconductor increases sharply as the value $\epsilon = \Delta$ is approached from above (see Fig. 3.1). The region of energies less than Δ is forbidden, which is precisely the physical meaning of a gap.

So far, we have been considering the behavior of an ideal electron gas. The effect of the static field of the lattice can be taken into account, for example, by introducing an effective electron mass. However, the picture gets more complicated when we try to include the lattice vibrations and the resulting electron–phonon interaction. This effect becomes especially relevant when the electron–lattice interaction is not weak (this situation will be discussed in more detail later; see Chapter 6). In this case, one introduces the so-called order parameter $\Delta(\epsilon)$, which should not be confused with the energy gap (the gap is defined as the root of the equation $\epsilon = \Delta(-i\epsilon)$.

The special form of the superconducting density of states described above leads to the following expression for the conductivity of the N–I–S tunneling junction:

$$\sigma_S = \sigma_N \nu_T(V) \tag{3.7}$$

The quantity $\nu_T(\omega) = \mathrm{Re}\{\omega[\omega^2 - \Delta^2(\omega)]^{-1/2}\}$ is called the tunneling density of states; it increases rapidly as ω approaches the energy gap Δ_0. The factor σ_N is the conductivity of the corresponding normal junction [see Eq. (3.6)]. This description can be generalized to the case of an S–I–S tunneling junction.

This analysis shows that by measuring the tunneling current, one can determine the magnitude of the energy gap. Indeed, when the applied potential approaches the value of the gap, Δ_0, the conductivity of the junction rises sharply. This corresponds to the simple physical picture described at the beginning of this chapter. The ability of tunneling experiments to measure the energy gap is very important.

This does not exhaust the role of tunneling in the physics of superconductivity. A more detailed analysis of the junction conductivity allows one to reconstruct the order parameter, which in turn provides unique information about the mechanism of superconductivity and the interaction strength. We will discuss this aspect of tunneling spectroscopy in Chapter 6, which is devoted to strong-coupling superconductivity.

One particular aspect of the physics of tunneling in superconductors, namely, the Josephson effect, will also be discussed separately (see Chapter 4).

ULTRASOUND ATTENUATION

Ultrasound attenuation in superconductors was one of the first phenomena to be studied both experimentally and theoretically. This is due to the great wealth of information about a superconductor's spectrum that can be obtained with this method.

Let us consider absorption of high-frequency sound waves. This process can be viewed as direct absorption of sound quanta by electronic excitations (it is analogous to Landau damping of plasma waves).

Ultrasound attenuation in superconductors is described by the following remarkably simple formula:

$$\frac{\gamma_s}{\gamma_n} = \frac{2}{e^{\Delta/T} + 1} \tag{3.8}$$

Here γ_s and γ_n are the absorption coefficients in the normal and superconducting states, respectively. Equation (3.8) is valid if $\hbar\omega < 2\Delta$, where ω is the sound frequency.

Equation (3.8) shows that superconductors absorb sound more weakly than normal metals. At absolute zero, $\gamma_s = 0$, that is, there is no absorption at all. Physically, this is due to the presence of the energy gap. A sound quantum which has low energy cannot excite the electronic system. At a finite temperature, sound is absorbed primarily by the "normal" component of the electron fluid, but still the presence of the energy gap makes the absorption weaker than in a normal metal, all the way up to the critical temperature.

Near $T = 0$ K, the absorption is characterized by an exponential dependence $\gamma_s \sim e^{-\Delta/T}$, which is due to the exponentially small number of excitations. This dependence is analogous to the temperature dependence of the electronic heat capacity of superconductors at $T \to 0$.

The formula in Eq. (3.8) was used in the following experiment. The ratio γ_s/γ_n was measured at different temperatures, and the formula was employed to calculate the energy gap, thus establishing the temperature dependence of the latter. The experimental results turned out to be undistinguishable from the theoretical curve $\Delta(T)$ (see Fig. 2.2). The excellent agreement confirms the validity of the principal ideas of the theory of superconductivity.

Ultrasonic measurements also provide information about gap anisotropy. We will discuss this question later in this chapter.

A few remarks should be made about the difference between the velocity of sound in superconductors as compared with normal metals. In ordinary superconductors, this difference is small, usually on the order of 10^{-6} (for lead, $\Delta u/u \approx 10^{-4}$). We conclude that the crystal lattice is not altered significantly by the superconducting transition. One can prove that the reason lies in the smallness of the quantity $[(\Delta(0)/E_F]^2$. Note that the situation is different in the new high-temperature superconductors (see Chapter 13).

ENERGY GAP ANISOTROPY

The energy gap, which is one of the main parameters characterizing the superconducting state, depends not only on the temperature, but also on the spatial direction. That is, $\Delta \equiv \Delta(\mathbf{p},T)$, where $\mathbf{p}$ is the electron momentum. This anisotropy is due to several factors. First of all, the electron–phonon interaction, which lies at the foundation of the BCS theory, depends on the mutual orientation of the electron and phonon momenta. Thus, the effective electron–electron interaction becomes anisotropic as well. Second, the dispersion of electron energies, $E(\mathbf{p})$, is important. Since the principal contribution comes from the states near the Fermi surface, one can say that the shape of the latter also affects the anisotropy of the energy gap. In the isotropic model, the effective interaction is assumed to be independent of the direction, and the dispersion law is taken to be quadratic: $E(p) = p^2/2m^*$ (m^* is the effective mass). This corresponds to a spherical Fermi surface. For real metals, both these assumptions are violated, although to different degrees in different materials. In addition, the vibrational spectrum of the lattice is also anisotropic. As a result, the energy gap becomes anisotropic; this should be reflected in various observations.

As the temperature increases, the gap becomes smaller and finally vanishes at $T = T_c$. It might seem that because of anisotropy, the gap may vanish sooner in some directions than in others. However, this kind of "extended" transition is not observed. It can be rigorously shown (Pokrovskii, 1961) on the basis of the BCS theory that the gap decreases uniformly in all directions and simultaneously vanishes for all of them at $T = T_c$.

How, then, does anisotropy manifest itself? Let us consider the behavior of the electronic heat capacity C^s_{el}. The presence of the energy gap leads to C^s_{el} vanishing exponentially as $T \rightarrow 0$ [see Eq. (5.1) below].

We have used the relation $\Delta(0) = 1.76T_c$ obtained in the isotropic BCS model. Anisotropy will change this dependence of the heat capacity. Clearly, those states on the Fermi surface which correspond to the smallest gap, $\Delta_{\min}$, assume increasingly important roles as the temperature decreases. As a result, we have $C^s_{el} \sim \exp(-\Delta_{min}/T)$. Since $\Delta_{min} < \Delta_{BCS} = 1.76T_c$, it is clear that gap anisotropy leads to a rate of decrease of C^s_{el} slower than that given by Eq. (3.9).

Similar considerations apply to sound attenuation, which is also proportional to $\exp(-\Delta_{min}/T)$ (cf. Eq. 3.8). Speaking of sound attenuation (this phenomenon is probably the most sensitive to anisotropy), we should note that due to the relation $u/v_F \ll 1$ (u is the speed of sound; for usual metals, this ratio is $\sim 10^{-3}$), attenuation is mostly caused by electrons moving almost perpendicular to the direction of wave propagation, that is, $v_F q = 0$ (q is the phonon momentum). This condition follows directly from conservation of energy and of momentum. They give $\mathbf{v}\cdot\mathbf{q} = \omega$. The condition $u \ll v_F$ finally leads to the equation $\mathbf{v}_F\mathbf{q} = 0$. Thus, sound attenuation at $T \to 0$ is governed by the minimum value of the gap, Δ_{min}, on the strip $\mathbf{v}_F\mathbf{q} = 0$ of the Fermi surface. By varying the direction of sound propagation, one can, in principle, reconstruct the function $\Delta(\mathbf{p})$.

These effects caused by the gap anisotropy, along with a number of others, have been observed experimentally, but they turned out to be not as large as one could have expected. Despite the huge anisotropy of the Fermi surfaces of many metals, the gaps do not differ too much from the isotropic BCS value. The success with which the isotropic model describes the experimental data may seem surprising at first glance.

The explanation for the small effect of anisotropy in ordinary superconductors was given by Anderson in 1959. According to the Anderson theorem, this fact is due to the large coherence length, $\xi_0 \simeq 10^{-4}$ cm (see above). The fact of the matter is that, in all real materials, electrons are scattered by impurities. This elastic process will scatter an electron from one section of the Fermi surface to another (by definition, all the states on the Fermi surface have the same energy, so such a process can take place under elastic scattering). In other words, the electronic states on the Fermi surface get all mixed up. Such collision processes would be unimportant if $\xi_0 \ll l$, where ξ_0 is the size of the Cooper pair and l is the mean free path. The large value of ξ_0 makes this condition impossible to satisfy even for very pure materials. As a result, the mixing of states takes place, leading to the gap becoming isotropic.

Let us note that the situation is completely different in the recently discovered high-temperature superconductors. We shall discuss this question in more detail in Chapter 13, devoted to the properties of the high-T_c oxides; we view the question of gap anisotropy as being one of the most fundamental in the physics of high-temperature superconductivity. Here we just remark that the small coherence length in the cuprates leads to the Anderson criterion $\xi_0 \ll l$ being satisfied, and the effects of anisotropy must be taken into account.

SUPERCONDUCTORS WITH OVERLAPPING BANDS

In the preceding section, we considered the influence of real crystal structure on the properties of superconductors. Anisotropies of the Fermi surface and of the interactions are directly related to crystal structure. In this section, we continue to discuss this general problem.

Let us consider a metal with two overlapping energy bands, "a" and "b" (see Fig. 3.4). We are limiting ourselves to the case of two bands for simplicity, although it is easy to generalize the analysis. We are thus dealing with two groups of electrons, each with its own density of states, degree of band filling, and so on. Each group is characterized by its own energy gap. This two-gap model was first introduced by Suhl *et al.* (1959) and was also considered by Moskalenko (1959). Later, a detailed analysis of multigap superconductivity was carried out by Geilikman *et al.* (in 1967 and 1972) (see Geilikman and Kresin, 1974).

Note that the case of overlapping energy bands can be considered as one of generalized anisotropy, that is, of a nonspherical Fermi surface. The overlapping bands correspond to topologically distinct sections of the Fermi surface, and clearly in this case the shape of the latter will be far from spherical. Nevertheless, we shall distinguish between the anisotropies of the energy gap and of the two-gap model. The gap can be anisotropic even if only one band is present, for example, because of the anisotropy of the dispersion law $\epsilon(\mathbf{p})$. When considering the multigap model, we wish to emphasize the presence of several different groups of electrons. Each group has its own energy gap, and each one of these, generally speaking, can be anisotropic. For simplicity, however, we will consider a system with two overlapping bands "a" and "b" and two isotropic gaps Δ_a and Δ_b.

In the two-band model, each band has its own set of Cooper pairs. That is to say, only electrons belonging to the same band pair up. This is reflected in the appearance of the two gaps, Δ_a and Δ_b. There is no pairing of electrons belonging to different bands; consequently, no pair potential is introduced. The reason for this is that only electrons with equal and opposite momenta (and spins) pair

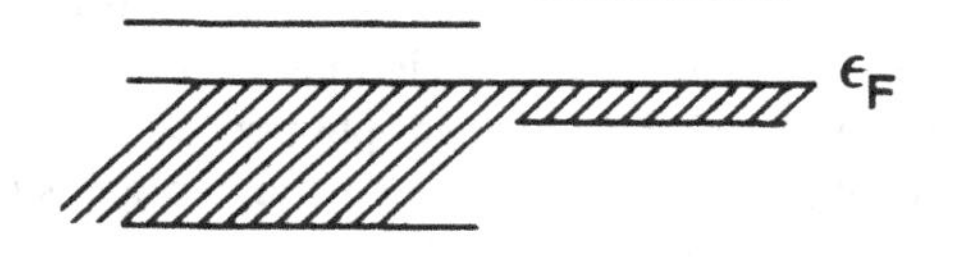

Figure 3.4. Overlapping bands.

up, whereas electrons located on the Fermi levels of different bands have markedly different momenta p_F^a and p_F^b (unless the Fermi surface has a very special shape).

In this regard, we should point out the following important and nontrivial feature of the two-band model of a superconductor. The absence of interband pairing does not mean that the two bands are two completely independent groups of carriers. In fact, electron pairs do make interband transitions. These transitions can be visualized as follows. One electron (say, from band "a") emits a phonon and makes a transition into band "b." Another "a"-electron absorbs this phonon and also goes into band "b," forming a Cooper pair with the first electron. In other words, two electrons from the same band can exchange a phonon, and the resulting attraction leads to their forming a bound pair in the other band. These acts of phonon emission and absorption represent so-called virtual (intermediate) processes. In such processes, as is well known, only the initial and final states are subject to energy conservation.

The process described here leads to the bands influencing each other. Whereas the usual BCS theory is characterized by one coupling constant, λ, describing the interelectron attraction, the two-band model has three constants, λ_a, λ_b, and λ_{ab} ($\lambda_{ba}=\lambda_{ab}\nu_a/\nu_b$). With increasing temperature, the energy gaps Δ_a and Δ_b decrease. Again, just as in the case of anisotropy, one could suspect that each one of the gaps would vanish at its own critical temperature, but, once again, this does not happen. The presence of the interaction constant leads to the system having a single critical temperature (in spite of there being two electron bands), equal to

$$T_c = 1.14\tilde{\Omega}e^{-1/\tilde{\lambda}};\ \tilde{\lambda} = \frac{1}{2}\,[\lambda_a + \lambda_b + \sqrt{(\lambda_b - \lambda_a)^2 + 4\lambda_{ab}\lambda_{ba}}] \tag{3.9}$$

At $T = T_c$, both energy gaps vanish. It is impossible to have a situation where one band has become normal and the other is still superconducting. Furthermore, one can show that the presence of interband coupling λ_{ab} is favorable for pairing and leads to a higher T_c.

In the BCS theory, the energy gap is related to T_c by the simple universal relation $\Delta(0) = 1.76T_c$. In the two-band model, there is no such simple relation. The quantities $\Delta_1(0)/T_c$ and $\Delta_2(0)/T_c$ depend on the constants λ_a, λ_b, and λ_{ab} and on the relationship between them. Deviations from the value $\Delta(0)/T_c|_{\mathrm{BCS}} = 1.76$ indicate the strength of the electron–phonon coupling (see Chapter 6). However, this statement is incorrect in the case of a two-gap system. In this case, the ratios $\Delta_i(0)/T_c$ can differ from the value given in Eq. (2.9) even for weak coupling. In fact, a theorem can be proven that shows that one gap is larger than the BCS value whereas the other one must be smaller, i.e., $\Delta_a > 1.176k_BT_c$ and $\Delta_b < 1.76k_BT_c$. Let us note, though, that the ratio Δ_a/Δ_b turns out to be the same at $T = 0$ K and at $T = T_c$.

Multigap superconductors have a number of interesting features. As we mentioned earlier, each band has its own set of pairs. The sizes of the pairs, $\xi_o^i = \hbar v_i/\pi\Delta_i(0)$, can be very different. This can lead to peculiar electrodynamics (see Chapter 4) because the local approximation may be valid for one band, and the nonlocal for another. As in the anisotropic case, the heat capacity of such a superconductor will decrease slower than in the usual case and will be proportional to $\exp(-\Delta_{\min}/T)$, where $\Delta_{\min}$ corresponds to the band with the smallest gap.

Overlapping energy bands are present in many metals, and one could expect noticeable deviations of the temperature dependences of the electron heat capacity, sound attenuation, and other properties from the predictions of the one-band model. This is not the case, however, and is explained in the same way as the small effect of gap anisotropy. Collisions with impurities lead to interband transitions and to a mixing of the states from different bands. As a result, the differences of the gaps are smeared out. Multigap structure can be observed only if the mean free path is much greater than the coherence length. This is realistic in the new high-temperature superconductors. The effects of multiband structure, just as those of anisotropy, are directly observed in these materials. We will return to this question in Chapter 13, devoted to the properties of the high-T_c oxides.

GAPLESS SUPERCONDUCTIVITY

Doping a superconductor with ordinary nonmagnetic impurities does not strongly affect such properties as the critical temperature and the energy gap. Only a very pure superconductor will suffer a small decrease in T_c (about 1%) when doped. This decrease comes to an end when the mean free path l becomes equal to the Cooper pair size ξ_0 (at which point the gap becomes isotropic, that is, independent of the direction.

A drastically different picture is observed when a superconductor is doped with magnetic impurities. Magnetic impurities possess a magnetic moment. This leads to an additional interaction between the impurities and the paired electrons. The two electrons in a Cooper pair have opposite spins and, consequently, opposite magnetic moments. Therefore, each one will be acted upon differently by an impurity atom. The latter will try to flip the magnetic moment of one of the electrons, thus destroying the bound state of the pair. Experimentally, one observes a marked change in the critical temperature and the energy gap whenever magnetic impurities are introduced. Even a small impurity concentration (a few percent) can lead to a complete destruction of the superconducting state.

In the presence of magnetic impurities, there can arise a peculiar "gapless" superconducting state (Abrikosov and Gor'kov, 1960; de Gennes and Sarma,

1963) (see de Gennes, 1966). It turns out that as one keeps adding the impurities, the gap decreases faster than T_c, and when the concentration reaches $n_0 = 0.91 n_{cr}$ (n_{cr} is the impurity concentration at which superconductivity disappears), the gap vanishes while the sample remains superconducting (for instance, there is still no electrical resistivity). The absence of a gap is observed experimentally in tunneling experiments. The observed current–voltage characteristic is ohmic: there is a nonzero tunneling current at arbitrarily small voltages, which means that there is no energy gap.

How does gapless superconductivity come about? In a qualitative way, this can be understood as follows. As we have already pointed out, magnetic impurities try to destroy electron bonding needed for superconductivity. At some impurity concentration n_0, some fraction of the Cooper pairs are broken up. Then even at $T = 0$ K, we have a picture similar to the two-fluid model. Because of the presence of the "free" electrons, created by the partial breakup of the Cooper pairs, the energy gap disappears (these electrons can absorb radiation of arbitrarily low frequency and occupy the states needed for the flow of tunneling current). The remaining Cooper pairs sustain resistanceless current flow.

The phenomenon of gapless superconductivity is possible not only in the presence of magnetic impurities. Any field which acts differently on the electrons in a pair and tries to destroy superconducting order can lead to a gapless superconducting state. For example, this state is found in thin films in the presence of an external magnetic field, in films carrying current close to critical, and so on.

The existence of gapless superconductivity shows that the presence of a gap is not a necessary condition for the appearance of superconductivity. The main thing is for there to be a bound electronic state of the pairs; this manifests itself by a nonvanishing order parameter. Such a state can display superconducting properties even in the absence of an energy gap.

MACROSCOPIC QUANTIZATION

QUANTIZATION OF MAGNETIC FLUX

One of the main principles of quantum physics is that a number of physical quantities, such as energy and momentum, are, under certain conditions, quantized, that is, can take on only a discrete set of values. However, until recently it was thought that quantization is relevant only in the microscopic world, that is, that it is characteristic of processes occurring in atoms, nuclei, and the like. Indeed, when studying macroscopic objects, we are dealing with an enormous number of particles, and, although the behavior of every single particle is governed by the laws of quantum physics, quantization of energy, momentum, and other physical quantities does not manifest itself on the whole due to the chaotic thermal motion of the particles. The thermal motion masks quantum regularities.

However, a number of phenomena, in particular, superconductivity, have shown that it is also possible to observe macroscopic quantization, that is, quantization of parameters which characterize macroscopic objects hundreds of thousands of times larger than the atomic dimensions. This is due to the fact that the electron system in a superconductor is highly correlated.

Suppose a persistent superconducting current is circulating in a metal ring. Its behavior is analogous to that of an electron in an atomic orbit. According to the quantum theory, a number of quantities describing the state of such an electron take on discrete values. However, whereas in studying the atom we are dealing with microscopic quantization which cannot be observed directly, superconductivity offers an example of quantization of a macroscopic quantity—electrical current. With the help of the superconducting ring, we are observing a quantum effect on an enormous scale.

It turns out that the current in a superconducting ring cannot be of an arbitrary magnitude and change continuously. A gigantic Bohr orbit arises for the entire electron group moving in the ring. Since the electrical current is related to the magnetic field, the latter will also be able to take on only a discrete set of values. Consequently, the magnetic flux $\Phi = \pi r^2 H$ through the ring will be quantized as well. In other words, $\Phi = N\Phi_0$, where N is an integer and Φ_0 is some minimum amount—the magnetic flux quantum. Magnetic flux is a macroscopic quantity, and the possibility of its quantization signifies a transition to a gigantic quantization scale, compared to the atomic one.

Let us now determine the magnitude of the magnetic flux quantum. We apply the Bohr momentum quantization condition to the electrons moving around the ring:

$$mvr = N\hbar \tag{4.1}$$

Here r is not the radius of an electron orbit in an atom, but the radius of the ring which contains the superconducting current. In this way, the scale of quantization has been increased by many orders of magnitude. Since the ring radius is a fixed quantity, the above condition can be taken as the momentum ($p = mv$) quantization rule. Momentum quantization implies that the current and, consequently, the magnetic flux are quantized as well. Let us find the relation between p and Φ.

The energy of a current I flowing in a loop of induction L is equal to $E = LI^2/2c^2$, and the magnetic flux is $\Phi = LI/c$. Therefore, $E = I\Phi/2c$. The current set up in the ring by n electrons moving with velocity v is $I = nve/2\pi r$. We thus arrive at the following formula for the energy:

$$E = \frac{\Phi}{2\pi}\frac{ne}{rc}\cdot\frac{v}{2} \tag{4.2}$$

On the other hand, the energy of n electrons moving around the ring with velocity v is equal to $E = nmv^2/2 = npv/2$. Comparing these two expressions for the energy, we arrive at the following formula for the momentum: $p = \Phi e/2\pi rc$. Since the electrons in a superconductor are paired up, the momentum of an electron pair is $p = \Phi e/\pi rc$. Substituting this into the quantization condition (Eq. 4.1), we obtain

$$\Phi = N\Phi_0 \tag{4.3}$$

where $\Phi_0 = hc/2e$ and $N = 1,2,3, \ldots$. Thus, the magnetic flux quantum is equal to $\Phi_0 = hc/2e$.

*Another method of calculating Φ_0 is not so elementary, but is more rigorous and compact. We start with the Bohr quantization condition

$$\oint \mathbf{p}_s\, d\mathbf{l} = nh \tag{4.4}$$

The momentum of an electron pair moving in a magnetic field is $\mathbf{p}_s = 2(m\mathbf{v} + e\mathbf{A}/c)$. Substituting this into Eq. (4.4), we find

$$\frac{2m}{ne} \oint \mathbf{j}_s \, d\mathbf{l} + 2 \frac{e}{c} \oint \mathbf{A} \, d\mathbf{l} = Nh \tag{4.5}$$

We have used the relation $\mathbf{j}s = ne\mathbf{v}_s$ between the superconducting current $\mathbf{j}_s$ and the velocity $\mathbf{v}_s$. Changing the integral with the vector potential $\mathbf{A}$ into one over the area surrounded by the path (remembering that $\mathbf{H} = \text{curl } \mathbf{A}$), we arrive at the following formula:

$$\frac{mc}{ne^2} \int_C \mathbf{j}_s \, d\mathbf{l} + \phi = N \frac{hc}{2e} \tag{4.6}$$

where $\Phi = \int \mathbf{H} d\mathbf{S}$ is the magnetic flux.

F. London called the quantity on the left-hand side of this equation a "fluxoid." If the path C lies away from the edge of the opening, at a distance greater than the penetration depth (we are assuming that $d \gg \delta$ and $r \gg d$ where d is the thickness of the superconducting ring and r is the radius of the opening), then along the path of integration $j_s = 0$ (recall that the superconducting current flows in a surface layer of thickness $\sim\delta$), and we arrive at Eq. (4.3).

As we have discovered, the magnitude of the magnetic flux quantum is $\Phi_0 = hc/2e = 2 \times 10^{-7}$ G-cm^2. Is this a large or a small quantity? Imagine a thin cylinder 0.1 mm thick. If the magnetic field inside this cylinder corresponds to one magnetic flux quantum Φ_0, it will equal approximately 1% of the Earth's magnetic field. Thus, the quantum corresponds to a macroscopic value of the magnetic field.

The flux quantum can be measured experimentally. One such experiment (Deaver and Fairbanks, 1961) involved exciting superconducting current in a small tin pipe 1 cm long, and with an inner diameter of just 1.5×10^{-3} cm. Owing to the extremely small diameter, a noticeable magnetic field was required to create one flux quantum $\Phi_0 = H\pi R^2$. It equaled 0.1 G, which is only five times less than the Earth's magnetic field. This field was measured by coils placed near the ends of the tube, which was oscillating along its axis. It was discovered that, indeed, the magnetic flux can take on only a discrete set of values and, as a result, can change only in jumps.

Magnetic flux quantization also can be directly observed in Josephson effect experiments (see next section). This involves turning on a magnetic field parallel to the plane of the plates which make up the Josephson junction. The field will affect the phase relationships and make the junction current spatially inhomogeneous. At some values of the field, the current will vanish. The values are quite definite and correspond to there being an integral number of magnetic flux quanta in the barrier layer.

Flux quantization was predicted by F. London in 1950, even before the appearance of the theory of superconductivity. However, he obtained an expression for the flux quantum Φ_0 that was twice as large as that derived above. At that time, nothing was known about electron pairs, and so London did not introduce the factor of 2 into the expression for the electron momentum. Experiment has confirmed the proposition that superconductors contain bound electron pairs.

It is curious that the experiment described here allows, through the use of the equation $\Phi_0 = \hbar c/2e$, the value of the Planck constant to be determined. As is well known, this is usually done only with the help of microscopic atomic methods. The phenomenon of superconductivity, being intrinsically quantum, makes it possible to determine the Planck constant by observing purely macroscopic phenomena.

THE JOSEPHSON EFFECT

In this section, we will describe another effect which exhibits the highly correlated character of the superconducting state. It is the Josephson effect, very interesting from the points of view of both basic science and numerous applications.

In 1962, a young theoretical physicist, Brian Josephson, predicted the possibility of observing two very unusual effects. He described his ideas in a two-page paper in *Physics Letters*. Reading this paper, one might get the impression that it is of purely academic interest. However, these effects were discovered soon afterwards, and now they form the foundation of superconducting electronics. It would be difficult to name another example where the time elapsed between basic physics and applications was so short.

There are two Josephson effects: stationary and nonstationary (also known as the dc and ac effects). The dc effect is that direct superconducting current can flow through a tunnel junction with no applied potential difference.

The discovery of the Josephson effect was of great importance not only in low-temperature physics. It provides insight into all of quantum physics. To explain better what we have in mind, let us use the well-known quantum-mechanical expression for the current density $j = (ie\hbar/2m)(\psi\nabla\psi^* - \psi^*\nabla\psi)$ and *remember that the wave function is, in general, a complex quantity:* $\psi = |\psi|\, e^{i\phi}$. Then we easily find that $j \sim \nabla\phi$. In real metals, in the absence of an external field, there is no macroscopic current flow; this is due to the fact that different electrons have random phases, and the current density averages to zero.

Superconductors, on the other hand, are characterized by electron ordering, or phase coherence. All electron pairs in a single superconductor have the same phase, and so there is no current in this case either: $\Delta\phi = 0$. However, if we form a tunnel junction consisting of two different superconductors, then a current will flow without any applied voltage. This current (the Josephson current) is determined by the phase difference $\phi = \phi_1 - \phi_2$; its density is given by the following simple expression: $j = j_0 \sin \phi$.

This effect, then, is governed by a fundamental quantum-mechanical quantity: the phase of a wave function. For the first time, there was an experiment in which a macroscopic quantity (the electric current) was directly determined by the phase of a wave function.

The dc Josephson effect was observed soon after its theoretical prediction (Anderson and Rowell, 1963). Figure 4.1 shows the current–voltage characteristic of a Josephson junction. If the junction is hooked up to a current source, the voltage across the barrier will be zero until the current exceeds its maximum value. Above this threshold, the junction exhibits ordinary single-particle tunneling.

The amplitude of the Josephson current depends on temperature. For an S–I–S junction (I is an insulator, surrounded by two identical superconductors S), this dependence is described by the following simple formula (Ambegaokar and Baratoff, 1963) (see Barone and Paterno, 1982):

$$j_0 = \frac{\pi\Delta}{2eR_N} \tanh \frac{\Delta}{2T} \tag{4.7}$$

Here R_N is the junction resistance in the normal state. The current is maximum at $T = 0$ K; then $j_0 = [\pi\Delta(0)]/(2eR_N)$. As $T \to T_c$, the amplitude of the current decreases, so that $j_0 \sim \Delta^2 \sim (T_c - T)$.

So far, we have been discussing the dc Josephson effect. It consists of a dc current flowing through a junction with no external voltage. This current is due to Cooper pair tunneling through a potential barrier.

What will happen if we do apply a voltage to a Josephson junction? This will result in the ac Josephson effect. It turns out that if we apply a *constant* voltage, then an *alternating* superconducting current will flow across the junc-

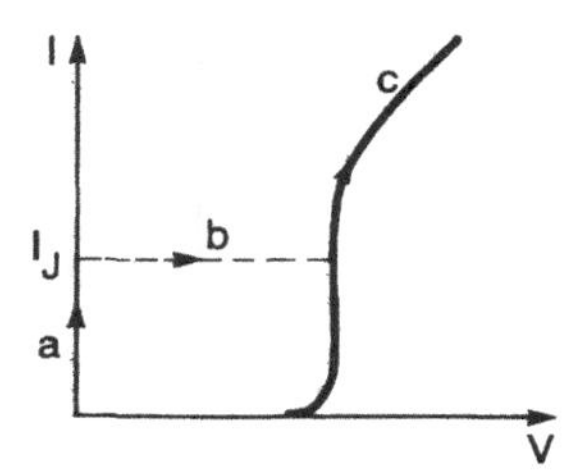

Figure 4.1. Current–voltage characteristic of a Josephson junction.

tion. This current will emit electromagnetic waves, just like any ac circuit, and these waves can be detected experimentally. This effect was observed for the first time in 1965 (Yanson *et al.*, 1965) (see Kulik and Yanson, 1972).

With a voltage U applied to the junction, bound electron pairs will cross the dielectric layer and acquire an energy equal to $2eU$. In a normal metal, this energy would be needed to overcome the resistance. No energy is expended when a superconducting current is flowing, so the portion $2eU$ acquired by the electron pair will be radiated as a light quantum of frequency $\omega = 2eU/\hbar$. This radiation is observed experimentally.

The expression for the frequency involves twice the electron charge. This is due to pairing. Conversely, the fact that one observes radiation of frequency $\omega = 2eU/\hbar$ can be viewed as an experimental confirmation of the theoretical concept of pairing in superconductors.

Thus, the Josephson effect allows the production of alternating current by means of a constant voltage. An applied voltage of 1 mV will result in a frequency $\omega = 4.85 \times 10^{11}\ \text{s}^{-1}$, which lies in the far infrared. In an indirect way, this radiation may be detected by the appearance of equidistant steps on the voltage–current characteristic if the junction is placed in a microwave cavity. The steps appear at voltages $V_{0n} = (nh/2e)\omega$, where ω is the frequency of the applied signal (Shapiro, 1963) (see Barone and Paterno, 1982).

The behavior of a Josephson junction in an external magnetic field deserves special mention. It turns out that the current is a nonmonotonic function of the applied field (see Fig. 4.2). This quantum interference effect is caused by the field

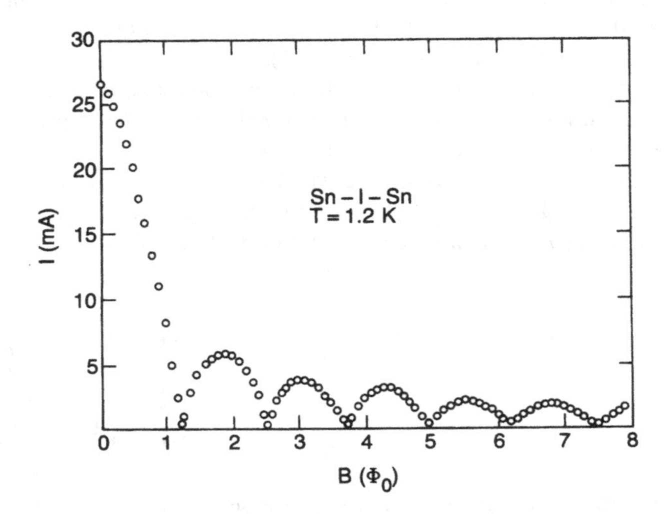

Figure 4.2. Josephson current as a function of magnetic field in a Sn–*I*–Sn junction.

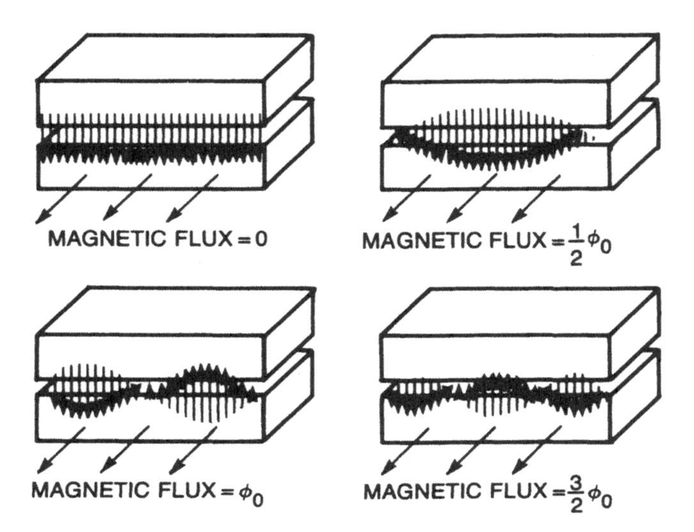

Figure 4.3. Josephson junction in magnetic field.

changing the phase of the wave function. As a result, the sign of the Josephson current depends on the field strength. Consequently, at some values of the external field, the current will vanish (see Fig. 4.3). The dependence of the amplitude of the current on the magnetic field is given by

$$j_M = j_0 \frac{\sin(\pi\phi/\phi_0)}{\pi\phi/\phi_0} \tag{4.8}$$

where Φ is the magnetic flux through the junction and Φ_0 is the flux quantum which we introduced in the preceding section. When the flux is zero or a multiple of the flux quantum, that is, $\Phi = n\Phi_0$, $n = 0,1,2, \ldots$, the Josephson current will vanish. Figure 4.2 illustrates these oscillations.

The Josephson effect plays a special role in superconducting applications. We shall describe this aspect in detail in Chapter 12.

5

THERMAL AND ELECTROMAGNETIC PROPERTIES

HEAT CAPACITY

The heat capacity of a metal is made up of the electronic and lattice contributions: $c = c_{el} + c_{lat}$. In usual superconductors, the superconducting transition has practically no effect on the lattice. As is known from the theory of normal metals, at low temperatures the dependence of the lattice heat capacity on temperature is given by $c_{lat} \sim T^3$. The same dependence characterizes c_{lat}^s in the superconducting state. In contrast, the electronic contribution changes drastically following the transition. We mentioned earlier (Chapter 1) that C_{el}^s is observed to decay exponentially as $T \rightarrow 0$. This behavior is explained by the theory of superconductivity as follows.

The number of quasiparticle excitations is given by the distribution function in Eq. (2.4). We have already pointed out that this number is less than the number of electrons in a normal metal and that it decreases exponentially with temperature.

The entropy of the system is determined by the quasiparticles (the entropy of the "normal" component, i.e., of the Cooper pair condensate, is equal to zero). From Eq. (2.4) it follows that the entropy, and consequently the heat capacity, decrease exponentially as $T \rightarrow 0$. To be more precise, for $T \ll T_c$, the heat capacity is given by

$$\frac{C_{el}^s(T)}{C_{el}^n(T_c)} = \frac{3\sqrt{2}}{\pi^{3/2}} \frac{\Delta(0)}{T_c} \left(\frac{\Delta(0)}{T} \right)^{3/2} e^{-\Delta(0)/T} \tag{5.1}$$

Thus, the presence of an energy gap leads to the electronic heat capacity behaving in a way radically different from the normal metal result, $c_{\text{el}}^{n} \sim T$.

The normal–superconducting transition is a second-order phase transition accompanied by a jump in the heat capacity. Since the lattice heat capacity does not change during the transition, it is clearly the electronic contribution that is responsible for the observed jump (Fig. 1.6). In the BCS theory, this jump is given by

$$\beta \equiv \left.\frac{C_s - C_n}{C_n}\right|_{T_c} = 1.43 \tag{5.2}$$

Note that in this case the universality of the BCS model is again apparent. This universality, just as in the relation between $\Delta(0)$ and T_c (Eq. 2.9) and many other relations that will be mentioned in this chapter, is due to the weak-electron–phonon-coupling approximation. Whereas the heat capacity jump is a universal property of all superconductors, its magnitude varies. Particularly striking deviations were observed in Pb and Hg; for example, in Pb, $\beta = 2.4$. For this reason, these superconductors were at first called "anamalous." Later, it became clear that these deviations from the BCS theory are due to strong coupling effects which are present in many superconductors. This subject will be discussed in Chapter 6.

THERMAL CONDUCTIVITY OF SUPERCONDUCTORS

Heat transfer in superconductors is characterized by some rather distinctive features. In the presence of a temperature gradient, dT/dx, a metal is not in a state of thermal equilibrium. There arises a heat flux Q in the sample, proportional to the temperature gradient present. The two are related by $Q = -\kappa dT/dx$, where κ is the thermal conductivity, which depends on the material and on temperature. The dependence $\kappa(T)$ has been well studied experimentally, and a lot of data are available for both normal and superconducting materials.

One has to distinguish between the thermal conductivity due to the motion of electrons and that of the crystal lattice. The fact that the lattice can contribute to the flow of thermal energy makes the phenomenon of heat conduction more complicated than electrical conduction. The latter represents transport of electric charge and consequently has only an electronic component.

The thermal conductivity κ can be written as a sum: $\kappa = \kappa_{\text{el}} + \kappa_{\text{lat}}$. The electronic thermal conductivity would be infinite if the electrons were not scattered by the thermal vibrations of the lattice, by impurities, and by other electrons. This statement is analogous to the fact that the electric conductivity would be infinite in the absence of collisions. There are, therefore, several mechanisms of thermal conductivity, related to the different kinds of scattering. The total electronic thermal conductivity is calculated according to the rule

$$\frac{1}{\kappa_{el}} \simeq \frac{1}{\kappa_{el\text{-}lat}} + \frac{1}{\kappa_{el\text{-}imp}} + \frac{1}{\kappa_{el\text{-}el}} \tag{5.3}$$

where the subscripts el–lat, el–imp, and el–el refer to electron–lattice, electron–impurity, and electron–electron scattering. This is analogous to the well-known law for the electrical resistance of a system of parallel conductors.

There are also several mechanisms determining the lattice thermal conductivity, namely,

$$\frac{1}{\kappa_{lat}} \simeq \frac{1}{\kappa_{lat\text{-}el}} + \frac{1}{\kappa_{lat\text{-}imp}} + \frac{1}{\kappa_{lat\text{-}lat}} \tag{5.4}$$

Thus, there are six different relaxation mechanisms. Each one depends in its own way on the temperature, electron concentration, etc. Each contribution is determined by the solution of the corresponding equation. For instance, the $\kappa_{el\text{-}imp}$ term is calculated from the equation $\kappa_{el\text{-}imp} = \int Ef\, d\tau/\nabla T$, where the integration is over all electronic states, and f is the electronic distribution function determined from the Boltzmann kinetic equation ($f = f_0 + f', f' \sim \nabla T$) for the case of impurity scattering.

In normal metals, the major role is played by the electronic thermal conductivity, which depends on temperature in the following way:

$$\kappa_{el}^{-1} = aT^2 + \frac{b}{T} \tag{5.5}$$

where a and b are constants.

The terms on the right-hand side correspond to $\kappa^{-1}{}_{el-lat}$ and $\kappa^{-1}{}_{el-imp}$ in Eq. (5.3). The κ_{el-el} term does not usually play an important role.

For a qualitative estimate, we can use the familiar formula from kinetic theory

$$\kappa = \tfrac{1}{3}\, lvc \tag{5.6}$$

where $l = v\tau$ is the mean free path (τ is the relaxation time), and c is the heat capacity. For example, for the scattering of electrons by impurities, we can write $\kappa_{el-imp} = \tfrac{1}{3} l_{el-imp} v c_{el}$. Here, l_{el-imp} is the temperature-independent electron mean free path, having to do with the collisions of the electrons with impurities. Since $c_{el} \sim T$, the temperature dependence of the corresponding thermal conductivity is $\kappa \sim T$ [see Eq. (5.5)].

When the impurity concentration is sufficiently large, the mean free path l_{el-imp} becomes small. This decreases κ_{el}, and the contribution of lattice conductivity becomes noticeable (although still small). At low temperatures, $\kappa_{lat} \simeq \kappa \sim T^2$. The lattice contribution can be determined with the help of the Wiedemann–Franz law, $\kappa_{el} = \frac{3}{2}(k_B/e)T\sigma$, and $\kappa_{lat} \simeq \kappa - \kappa_{el}$, where σ is the usual electrical conductivity.

In pure metals, the principal role is played by the thermal flow of electrons, which are scattered by phonons. In this case, the thermal conductivity increases with decreasing temperature ($\kappa_{el-lat} \sim T^{-2}$); this is due to the fact that the number of phonons decreases with decreasing temperature, and the electron mean free path grows.

The picture of thermal conductivity given above is analyzed in detail in the quantum theory of metals and is in good agreement with experiment. In superconductors, all the aforementioned mechanisms of heat conduction are at work as well, but the temperature dependence of the effect turns out to be completely different.

For instance, Fig. 5.1 shows the temperature dependence of the thermal conductivity of pure lead in the superconducting (s) and normal (n) states (the normal state can be induced by placing the sample in a magnetic field). In the latter case, the thermal conductivity grows with decreasing temperature, while in superconductors, on the contrary, κ decreases (basically, exponentially) as the temperature drops.

The heat transfer picture is quite remarkable in superconducting alloys in which, for certain concentrations of components, the main role is played by the lattice conductivity (Fig. 5.2). In normal samples the thermal conductivity drops with decreasing temperature, but in a superconducting alloy, as the figure shows, the thermal flow increases; at some temperature, κ_{lat} reaches a maximum.

These experimental observations serve to remind us once again about the

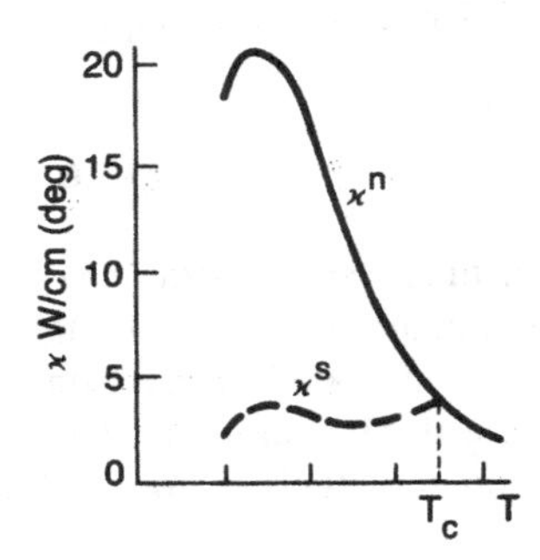

Figure 5.1. Thermal conductivity of lead in the superconducting and normal states.

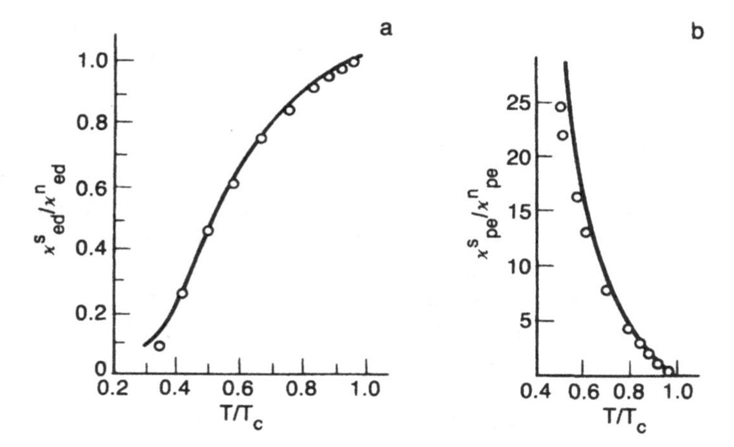

Figure 5.2. (a) Temperature dependence of κ^s_{ed}: solid line is the theoretical curve (Eq. 5.7); ○, experimental data for Tl. (b) Temperature dependence of κ^s_{pe}: solid line is the theoretical curve; ○, experimental data for an In–Tl alloy.

special character of the superconducting state, as evidenced by its physical properties, which differ in many respects from those of a normal metal.

The microscopic theory of superconductivity succeeds in describing the features of heat transfer in superconductors. In superconductors, the electronic component of the heat flux is determined only by the quasiparticle excitations [see Eqs. (2.1) and (2.4)], whose number decreases exponentially as the temperature is lowered. As far as the Cooper pair condensate is concerned, its entropy is zero, and it does not contribute to the energy transfer. In the language of the two-fluid model (see Chapter 2), this means that only the "normal" component transports heat.

The exponential decay of the number of excitations with decreasing temperature (below T_c) is the decisive factor. As a result, the electronic thermal conductivity in the superconducting state, κ^s_{el} also decreases exponentially. This can be seen from the relation (5.6): the heat capacity c^s_{el} decreases exponentially, and consequently so does κ^s_{el}. It is clear that considerations based on Eq. (5.6) are only qualitative. A detailed analysis carried out on the basis of the microscopic theory [by Geilikman in 1958 (see Geilikman and Kresin, 1974)] leads to the following result:

$$\frac{\kappa_s}{\kappa_n} = \frac{6}{\pi^2} T \left\{ \frac{\Delta^2/T}{e^{\Delta/T} + 1} + 2T \sum_{s=1,2\ldots}^{\infty} \frac{(-1)^{s+1}}{s^2} e^{-s\Delta/T} + 2\Delta \ln(1 + e^{-\Delta/T}) \right\} \quad (5.7)$$

Here κ_s and κ_n are the thermal conductivities in the superconducting and normal states, respectively. It is seen that the ratio κ_s/κ_n is a universal function of the temperature. Figure 5.2 shows that there is good agreement between experiment and theory.

The behavior of the lattice thermal conductivity also turns out to be rather unusual [see Geilikman and Kresin, 1974 and Bardeen and Schrieffer (in Gorter, 1964)]. In the quantum language, which must be used for analyzing phenomena at low temperatures, lattice heat conduction can be visualized as the transfer of energy by the sound quanta (phonons) which describe the vibrational state of the lattice. The energy of lattice vibrations decreases with temperature, which in a normal metal leads to the lattice thermal conductivity dropping as $T \rightarrow 0$. In the normal state ($T > T_c$), the contribution of κ_{lat} is small and has the temperature dependence $\kappa_{\text{lat}} \simeq \kappa_{\text{ph-el}} \sim T^2$.

The superconducting transition (in usual superconductors) does not change the state of the lattice. The number of phonons depends on the temperature in the same way as in a normal metal, and therefore it would seem that κ_{lat} should also decrease in the usual way. However, it is observed that when the superconducting transition takes place, κ_{lat} becomes a rapidly rising function as the temperature decreases (see Fig. 5.2). The origin of this exponential growth can be found from the relation $\kappa_{\text{lat}} \sim l_{\text{ph}}$. In this temperature region (just as for $T > T_c$), the lattice thermal conductivity is mainly governed by the scattering of phonons by electronic excitations. The exponential decay of the number of such excitations at $T < T_c$ results in the corresponding growth of the phonon mean free path $l \cong l_{\text{ph-el}}$ and in the observed growth of the lattice thermal conductivity. In other words, the decrease in the number of phonons is accompanied by the far more effective growth of their mean free path, leading to the increase in κ_{lat}. This has been observed in superconducting alloys. We shall not write out the analytical expression for $\kappa_{\text{ph-el}}$. Figure 5.2 demonstrates the good agreement between experiment and theory.

As described above, when the temperature is lowered below T_c, the lattice thermal conductivity increases, but the growth is not endless. It continues until the mean free path connected with the scattering by electrons becomes larger than that related to the scattering by impurities and crystal boundaries. Then the latter kinds of scattering begin to dominate and κ_{lat} decreases because of the drop in the number of phonons as $T \rightarrow 0$. As κ_{lat} switches from growth to decline, it displays a maximum which can be seen experimentally.

Thus, the superconducting state is distinctive in that the role of the lattice thermal conductivity is considerably enhanced. This is due to the opposite behavior of the electronic and lattice components at $T < T_c$. The former undergoes exponential decay, whereas κ_{lat}, on the contrary, exponentially increases.

On the whole, one has to distinguish three major types of heat conduction in superconductors. Pure and slightly doped superconductors show the first kind (Fig. 5.3a). In these, the electronic thermal conductivity plays the major role as T approaches T_c and in the intermediate temperature range. With a decrease in temperature, κ declines on the whole exponentially. Figure 5.2a shows the temperature dependence for κ for thallium (first type of conductivity). As $T \to 0$, the lattice thermal conductivity κ_{ph-imp} begins to dominate, and κ approaches zero according to a power law.

Superconductors with a high impurity concentration show the second type of thermal conductivity (Fig. 5.3b). The thermal conductivity of strongly doped superconductors, as opposed to those of the first kind, increases just below T_c as the temperature is decreased. In this case, the lattice thermal conductivity is the major factor. A characteristic feature of this case is the presence of a maximum conductivity; after passing through the maximum, $\kappa(T)$ approaches zero in the manner of κ_{ph-imp}. The Pb–Bi alloy (Fig. 1.7) is an example of a superconductor with thermal conductivity of the second type.

Finally, the third type of thermal conductivity (Fig. 5.3c), which is encountered in moderately doped superconductors, is a combination of the first (for $T \cong T_c$) and the second type (in the intermediate temperature range). The data for Pb and Sn specimens suggest that these are examples of superconductors of the intermediate type. In these, κ_{el} is the major factor near T_c, and therefore the thermal conductivity initially falls with decreasing temperature. However, subsequently κ_{lat} becomes greater than κ_{el}, and in the intermediate region these superconductors behave like substances with the second type of thermal conductivity.

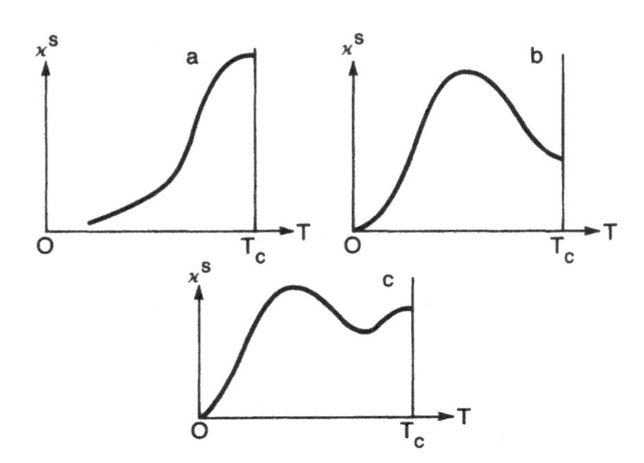

Figure 5.3. The basic types of thermal conductivity: (a) pure superconductors; (b) superconductors with a high impurity concentration; (c) the intermediate case.

ANISOTROPIC THERMAL EFFECT

A number of distinctive thermal effects can be observed in superconductors. We shall not attempt here to present a detailed discussion, but will describe just one of them, the so-called anisotropic thermal effect. The possibility of this effect was first pointed out by Ginzburg in 1944.

The concept of this effect is as follows. Consider a superconducting crystal with an applied temperature gradient ∇T. The latter gives rise to an electronic heat flow proportional to the gradient. In other words, there will be created a flow of electrons carrying thermal energy. However, this flow of electrons will transfer not only energy, but electric charge as well. That is, there will appear an electric current $\mathbf{j}_n$. In a superconductor, this transfer will be carried out by the "normal" component, that is, by the quasiparticle thermal excitations. The charge flow will lead to the appearance of an electric field. This, in turn, will result in a compensating current of Cooper pairs, $\mathbf{j}_s$ (the "superconducting" component), so that the total electric field will vanish.

In an isotropic crystal, where the only preferred direction is that of the temperature gradient, ∇T, the currents $\mathbf{j}_s$ and $\mathbf{j}_n$ are equal and opposite, so that the total current vanishes: $\mathbf{j}_s + \mathbf{j}_n = 0$.

A different situation arises in an anisotropic crystal, where, in addition to T, there exists another distinctive direction, namely, the crystal axis. In this case, the two currents do not necessarily cancel, and it is possible to end up with a nonzero resultant current and the associated magnetic moment.

Thus, the anisotropic thermal effect consists in the following: when a temperature gradient is applied to an anisotropic crystal, a magnetic field appears.

An analysis carried out on the basis of the microscopic theory leads to the following estimate for this field:

$$H_Z \cong \frac{4\pi}{c} (\nabla T)^2 \, \delta_{L0}^2 \, b_n \, T_c^{-1} \, (1 - T/T_c)^{-1} \tag{5.8}$$

where δ_{L0} is the London penetration depth at $T = 0$ K, and b_n is the thermoelectric coefficient in the normal metal.

The effect is most notable near T_c (although one should note that the above expression is not valid in the immediate vicinity of T_c). The effect is proportional to $(\nabla T)^2$ and, in addition, depends on the ratio $\eta = T_c/E_F$, $(b_n \sim \eta)$, which is very small in usual superconductors. This small value is what makes the observation of the anisotropic thermal effect difficult. We wish to stress this dependence on η because in the new high-temperature superconductors this quantity is not so small, and the effect becomes observable (see Chapter 13). Organic superconductors also have very small Fermi energy and are anisotropic (see Chapter 9). Although T_c is not so large ($T_c \sim 12$ K) we think the organic materials are the best to look for this effect.

ELECTROMAGNETIC PROPERTIES OF SUPERCONDUCTORS. PENETRATION DEPTH

We have already talked about the peculiar electromagnetic properties displayed by superconductors. The absence of electrical resistance, the Meissner effect, the ability of the critical magnetic field to destroy superconductivity—all these features have long attracted physicists' attention. Back in 1935, the brothers F. and H. London put forth equations for the density of the superconducting current.

Consider the well-known quantum-mechanical expression for the current in the presence of a field:

$$\mathbf{j} = \frac{ie\hbar}{2m}\left[(\nabla\psi^*)\psi - \psi^*\nabla\psi\right] - \frac{e^2}{mc}\mathbf{A}\psi^*\psi \qquad (5.9)$$

The first term on the right-hand side is the so-called paramagnetic term, while the second term results in a diamagnetic contribution to the magnetic moment. The magnetic field enters into Eq. (5.9) directly: $\mathbf{A}$ is the vector potential of the field. However, the field also enters in an indirect fashion: switching on the field changes the wave function ψ (in metals, ψ is the many-particle wave function describing the entire electron system). This "deformation" of the wave function leads to the paramagnetic term being nonzero in such many-body systems as ordinary metals and molecules. In normal metals, the first and second terms in Eq. (5.9) cancel, leaving only the weak Landau diamagnetism.

The Londons assumed that superconductors behaved differently. That is, in the case of superconductors the wave function is "rigid"; because of the correlated nature of superconductors, it does not change in the presence of a magnetic field. As a result, the paramagnetic term vanishes. The superconducting current then becomes proportional to the vector potential, which is the distinctive feature of the London equation.

The modern theory of superconductivity treats the electromagnetic properties microscopically. The interelectron correlation does indeed lead to diamagnetism, but the detailed picture turned out to be more complicated.

As we have discussed earlier, the quantum-mechanical analysis has proven that there exists a gap in the energy spectrum. The electron system in a superconductor can be visualized as consisting of bound electron pairs, and the pair size ξ_0 is quite large ($\xi_0 \approx 10^{-4}$ cm).

How, then, do these features affect the electromagnetic properties of superconductors?

As is well known, direct current flow in normal metals obeys Ohm's law: $\mathbf{j} = \mathbf{E}/\rho$ ($\mathbf{j}$ is the current density, $\mathbf{E}$ is the electric field strength, and ρ is the resistivity).

The current density at each point is determined by the electric field strength at this very point. Such a relationship between two quantities is called "local."

Inside a superconducting sample, there is no resistance to constant current flow, and no electric field. As was described above, a surface current is set up when the sample is placed in an external magnetic field. The theory of superconductivity establishes the connection between the current and the external field. It turns out—and this is one of the most essential peculiarities of the electrodynamics of superconductors—that the relationship is, generally speaking, nonlocal. The superconducting current at some point M is not determined just by the field at this point, but depends on the field in some neighborhood of the point M. Physically, this is due to the fact that the electrons in a superconductor are spatially correlated with each other. If the field alters the state of one electron, the interelectron interaction will affect the behavior of another one. This fact is reflected in the nonlocal character of superconducting electrodynamics.

The current density is related to the field by the formula

$$\mathbf{j}(\mathbf{r},\mathbf{r}')=\int K(\mathbf{r},\mathbf{r}')A(\mathbf{r}')d\mathbf{r}' \tag{5.10}$$

where the kernel $K(\mathbf{r},\mathbf{r}')$ is a complex function derived in BCS theory which depends on temperature, energy gap, the Fermi velocity, etc.

It is important that the relation represented by Eq. (5.10) is integral and nonlocal. One has to take into account both terms in the expression (5.9) for the current; that is, the wave function is not completely rigid.

If the field does not vary strongly over distances of the order of ξ_0, the relationship between $\mathbf{j}$ and $\mathbf{A}$ becomes local:

$$\mathbf{j} = -\tilde{K}\mathbf{A} \tag{5.11}$$

Equation (5.11) is precisely the London equation. In this case, the field penetration depth into the superconductor is large compared with the Cooper pair size ξ_0. Superconductors of this type (they include, for instance, Pb and In) are called London type. In the opposite case, the relationship between the current and the field will be nonlocal. Samples of this kind (e.g., Al, Hg) are termed Pippard type after the British physicist who, prior to the appearance of the theory of superconductivity and using semi-intuitive considerations, proposed the idea of such a nonlocal relationship.

The penetration depth increases as the critical temperature is approached. Near T_c, all superconductors are of the London type. However, this may take place only within a very small temperature interval. For example, Al exists in the London phase only in the interval $\Delta T/T_c \approx 10^{-4}$ near T_c.

With the help of Eq. (5.11), it is easy to prove the existence of the Meissner effect (see Fig. 5.4). Indeed, consider the following system of two equations:

$$\mathbf{j} = -\tilde{K}\,\mathbf{A}, \qquad \text{curl}\,\mathbf{H} = \frac{4\pi}{c}\,\mathbf{j} \tag{5.12}$$

The first one is the London equation, and the second is one of the Maxwell equations. Taking the curl of both sides of the first equation, we obtain curl $\mathbf{j} = -\tilde{K}\mathbf{H}$. Doing the same with both sides of the Maxwell equation gives curl (curl $\mathbf{H}$) = $(4\pi/c)$curl $\mathbf{j}$, or $-\Delta\mathbf{H} = (4\pi/c)$curl $\mathbf{j}$ (we have used another Maxwell equation: div $\mathbf{H} = 0$).

Using the result for curl $\mathbf{j}$ obtained from the London equation, we finally arrive at the following differential equation describing the spatial dependence of the magnetic field:

$$\Delta\mathbf{H} - \delta^2\mathbf{H} = 0 \tag{5.13}$$

If we consider a plane superconductor–vacuum boundary, then for the half-space $x > 0$ occupied by the superconducting material, the solution of the above equation is

$$H\,(x) = H(0)\,\exp(-x/\delta) \tag{5.14}$$

where $\delta^2 = \tilde{K}^{-1}$. This solution describes an exponential decay of the field in the superconductor. In this way, we arrive at the Meissner effect.

Figure 5.5 shows the temperature dependence of δ, the penetration depth.

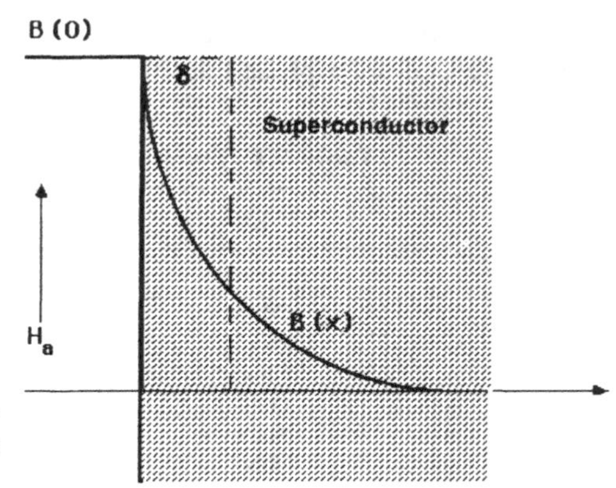

Figure 5.4. The penetration of the magnetic field into the superconducting sample.

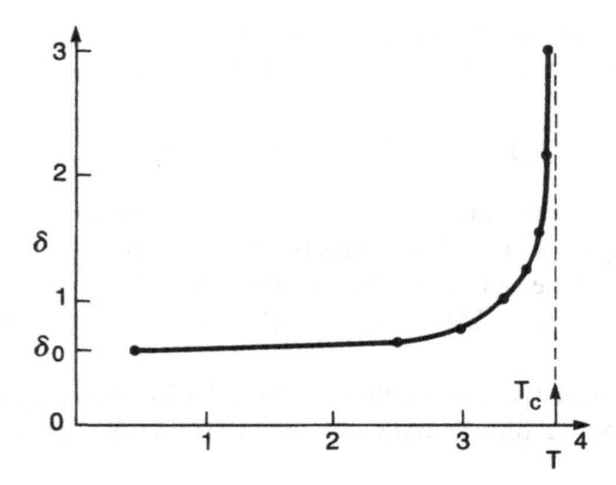

Figure 5.5. Penetration depth versus temperature.

Suppose we have a superconducting film thicker than the penetration depth at $T = 0$ K. Near T_c, the field may penetrate through the entire film, and one can say that the film has lost its anomalously large diamagnetic properties. However, this will be true only very close to T_c. As is seen from the figure, the penetration depth decreases very fast as T decreases. Unless specified otherwise, when talking about the penetration depth δ, we shall always mean the quantity δ(0), that is, the value at $T = 0$ K. The values of δ for some metals are given in Table 5.1.

THE GINZBURG–LANDAU THEORY

In 1950, Ginzburg and Landau created the phenomenological theory of superconductivity that bears their names. They based their work on Landau's theory of second-order phase transitions. The Ginzburg–Landau theory is valid

Table 5.1. Penetration Depths

Element	δ(0) Å
Al	500
Cd	1300
Pb	390
Nb	470
Tl	920
Nb_3Sn	650
NbN	2000
$Y_1Ba_2Cu_3O_7$	1500

for temperatures close to T_c and leads to a description of the behavior of superconductors in strong fields.

It is interesting to note that the Ginzburg–Landau theory, which is based on very nontrivial equations, was created before the microscopic BCS theory. After the latter was formulated, the validity of the Ginzburg–Landau theory was rigorously shown (Gor'kov, 1959). The only change that was made had to do with the magnitude of the effective charge e^* appearing in the theory. The authors had put it equal to the electron charge; the exact derivation led to $e^* = 2e$, which is a consequence of the formation of Cooper pairs.

The fact that the Ginzburg–Landau theory was formulated prior to the microscopic theory makes it an extraordinary example of the triumph of intuition and general physical principles.

*The equations of the theory have the following form:

$$\left\{ \frac{1}{4m} [\nabla_{\mathbf{r}} - 2ie\, \mathbf{A}(\mathbf{r})]^2 - \beta^{-1} \left[\frac{T_c - T}{T_c} - \frac{1}{n} |\psi(\mathbf{r})|^2 \right] \right\} \psi(\mathbf{r}) = 0 \qquad (5.15)$$

$$\mathbf{j}(\mathbf{r}) = -\frac{ie\hbar}{2m} [\psi^*(\mathbf{r})\nabla_{\mathbf{r}}\, \psi(\mathbf{r}) - \psi(\mathbf{r})\, \nabla_{\mathbf{r}}\psi^*(\mathbf{r})] - \frac{2e^2}{mc} \mathbf{A}(\mathbf{r})|\psi(\mathbf{r})|^2 \qquad (5.16)$$

Here n is the total electron density in the normal metal, and $\beta = [7\zeta(3)/6(\pi T_c)^2]E_F$, where E_F is the Fermi energy. Let us discuss the meaning of the function $\Delta(\mathbf{r})$. It is called the order parameter and is coordinate dependent. The quantity $\Delta(\mathbf{r})$ should not be confused with the energy gap; they become the same thing in a homogeneous medium where $\Delta(\mathbf{r}) = \Delta$ is independent of the coordinates. In principle, $\Delta(\mathbf{r})$ may be considered the wave function of a Cooper pair.

The major advantage of the Ginzburg–Landau theory is that it allows the study of spatially inhomogeneous systems, such as proximity systems, thin films, and others. It is most extensively used for analyzing the behavior of superconductors in an external magnetic field. Because of the spatial variation of the magnetic field, such systems are also spatially inhomogeneous. It is important that the theory describes superconductors in an arbitrary magnetic field, up to the critical value. Because of this, it is the basis of the analysis of the critical field and the critical current problems (see below).

The Ginzburg–Landau equations can be written in different ways. For example, they can be written in a dimensionless form. If this is done, there appears an important parameter, $\kappa = 0.96\delta_L/\xi_0$, where δ_L is the London penetration depth (note that near T_c all superconductors become London type) and ξ_0 is the coherence length. The form of the solution to the equations, and consequently the way superconductors behave, depends on the value of this parameter. There are two main kinds of superconductors: Type I ($\kappa < \sqrt{2}$), for which the penetration depth is less than the coherence length; and Type II ($\kappa > 1/\sqrt{2}$), where the reverse is true. We shall come back to this classification later, when we discuss the problem of the critical field. At this point, we shall only remark about the important effect of impurities on the electromagnetic properties of superconductors.

If a small amount of static (and nonmagnetic) impurities are introduced into a superconductor, this will have practically no effect on its thermodynamic properties and will not change its critical temperature. The electromagnetic properties are another matter. The processes of electron–impurity scattering which will change the electronic momenta are important and will affect the coherence length, ξ. The latter will become equal to $(\xi_0 l)^{1/2}$, where l is the mean free path. Thus, doping will lead to a decrease in ξ and a corresponding rise in κ. In this way, a pure Type I superconductor can, by doping, be turned into a Type II superconductor, resulting in a radical change of its behavior in an external magnetic field.

This mechanism is precisely the reason why most alloys are Type II superconductors, which accounts for their use in generating superstrong magnetic fields and currents.

CRITICAL FIELDS. THE MIXED STATE

The flow of superconducting currents does not generate heat. It is therefore very tempting to use them to obtain strong magnetic fields. However, it is clear that a superconducting coil cannot create fields stronger than the critical field H_c, since they would destroy the superconducting state. At first glance, it appears that there can be no hope for obtaining high critical fields. Indeed, let us make a simple estimate of the magnitude of H_c.

The Meissner effect of magnetic field expulsion will be taking place while the associated energy cost is offset by the more effective energy savings due to the metal being in the superconducting state. In sufficiently strong fields, it turns out to be energetically favorable to be not in the superconducting, but in the normal state with the magnetic field freely permeating the sample. The minimum value of the external field strength which destroys superconductivity is called the critical field. It can be determined from the equation

$$H_c^2/8\pi = W_b \tag{5.17}$$

where W_b is the energy it takes to destroy pairing. The bound state is formed not by all electrons, but only by those in the energy range Δ. The number of Cooper pairs is proportional to $\nu\Delta$ (ν is the number of electronic states per unit energy interval). To destroy a pair, it takes an amount of energy equal to the binding energy, that is, $\sim\Delta$. Consequently, $W_b \sim \nu\Delta^2$. From Eq. (5.17) we get the following estimate for the critical field:

$$H_c \sim \sqrt{\nu}\,\Delta \tag{5.18}$$

H_c is largest at $T = 0$ K, when

$$H_c(0) \sim \sqrt{\nu}\,\Delta(0) \tag{5.19}$$

For ordinary superconductors, the energy gap $\Delta(0)$ is simply related to the critical temperature [in the BCS theory, $\Delta(0) = 1.76T_c$], so that $H_c(0) \sim \nu^{1/2}\,T_c$. This estimate results in small values of H_c. For ordinary superconductors, $H_c \approx 10^2$–10^3 G. The high-temperature superconductors hace larger values of H_c, but even those are not sufficient for serious practical applications.

So the impression arises that superconductors can be used in magnets which would suffer no heat losses, but would be unable to produce very strong magnetic fields. However, this conclusion is premature, because there is another line of attack available.

On p. 59, we discussed the classification of superconductors into two types. The analysis described in the preceding paragraphs applies to Type I superconductors in which the Ginzburg–Landau parameter κ is less than $1/\sqrt{2}$. In other words, in these materials the penetration depth δ is less than the coherence length ξ_0. Type I superconductors are described by a simple phase diagram (Fig. 1.5) and do in fact have small critical fields.

The situation in Type II superconductors is completely different and more interesting. These compounds ($\kappa > 1/\sqrt{2}$, and the penetration depth exceeds the coherence length) can exist in the so-called mixed state. This state was analyzed by Abrikosov (in 1957) (see Abrikosov, 1988) on the basis of the Ginzburg–Landau theory.

The total expulsion of the magnetic field is due to absolute screening by the surface currents, which is energetically unfavorable. The energy cost would be reduced if the field were allowed to partially penetrate the superconductor. This is precisely the nature of the mixed state, in which the magnetic field is not completely expelled from the sample, but partially penetrates into the bulk, while the sample remains superconducting. This makes it possible to obtain strong magnetic fields without any electrical resistance and the associated power losses.

Superconductors are characterized by a certain surface energy. Its origin can be understood by the following line of reasoning. Figure 5.6 illustrates a normal phase region in equilibrium with a superconducting phase region. In order for the equilibrium to be sustained, inside the normal phase ($x > 0$) there must be a magnetic field equal to H_c, since $H_c^2/8\pi$ is the amount by which the phases differ in energy. The magnetic field penetrates into the superconductor's surface layer to the depth δ. Suppose that $\delta < \xi_0$. The order parameter $\Delta(r)$ decreases as we approach the normal region, with the length scale for the decrease on the order of ξ_0. The rapid attenuation of the magnetic field will lead in this case to

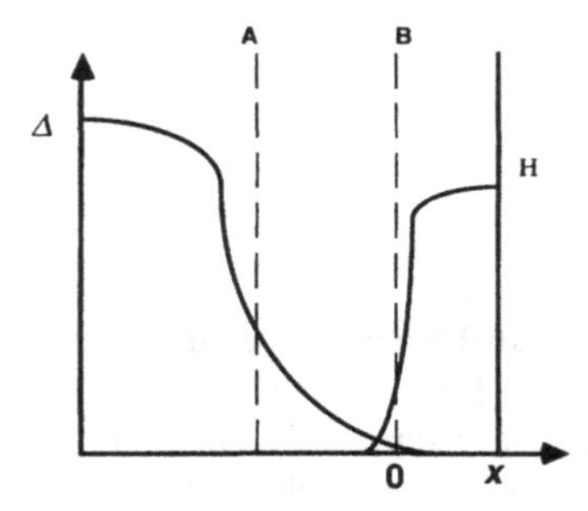

Figure 5.6. Coexistence of the normal and superconducting phases.

there being a region near the interface where both the order parameter and the field H almost vanish (region AB in the figure). In this region the pairing energy vanishes, but the field does not penetrate here. Thus, the screening of the field is not compensated by the gain in energy due to pairing. This results in the surface layer possessing some positive energy (surface energy).

In the opposite case when $\delta > \xi_0$ (this inequality is characteristic of Type II superconductors), the surface energy is negative. In this case, instead of the picture given in Fig. 5.6, there arises a region where the magnetic field coexists with Cooper pairing. In this region, both the order parameter $\Delta(\mathbf{r})$ and the magnetic field are finite. Such coexistence is energetically favorable and preserves the superconducting state. Thus, the magnitude of the parameter κ or, in other words, the relationship between the penetration depth and the coherence length plays an essential role in determining the critical field. Type I superconductors ($\delta < \xi_0$) have positive surface energies. All pure superconducting metals (except Nb) are of this type.

A pure metal can be turned into a Type II superconductor by doping. Electron–impurity collisions disrupt the pairing correlations. Recall that in usual superconductors the size of a pair is a large quantity ($\xi_0 \approx 10^{-4}$ cm $\gg a$, where a is the interatomic spacing, of order 10^{-8} cm), so that even small impurity concentrations can strongly affect the superconducting state. The case of negative surface energy is easily realizable in superconducting alloys. Instead of alloys, one can also use thin films of pure metals; in this case, the role of the mean free path will be played by the film thickness. It is interesting to point out in this connection that the critical field increases with decreasing film thickness.

The way in which a magnetic field penetrates into a Type II superconductor is very peculiar. As long as the field is weak, it is completely expelled. When a certain value H_{c1}, called the lower critical field, is reached, the field begins partially to penetrate into the superconductor. Under the influence of the Lorentz force, a fraction of the electrons (those with velocities perpendicular to the field lines) begin to move in a circle. This results in the appearance of the so-called

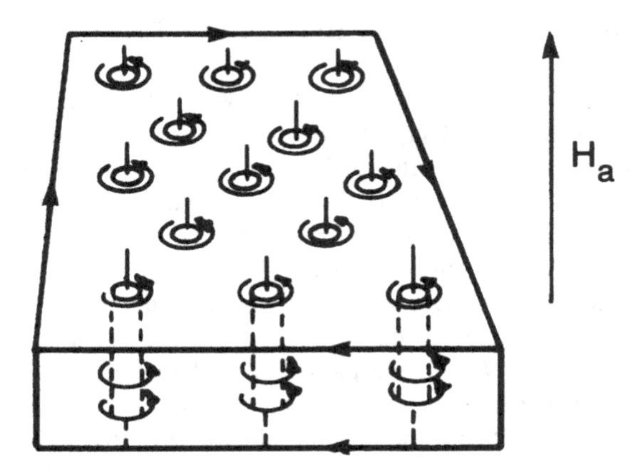

Figure 5.7. Mixed state.

vortex lines in the superconductor (Fig. 5.7). The superconducting electrons revolve around the vortex axis. This resembles the picture of the current in a superconducting ring discussed in Chapter 1. The closer to the vortex axis, the faster the electrons circulate. At some distance from the axis, the speed exceeds the critical value and superconductivity is destroyed. Therefore, the "core" of the vortex is in the normal state. There is a finite magnetic field inside the vortex.

Thus, the superconducting sample ends up perforated by lines of normal nonsuperconducting regions. The vortices form a triangular lattice. Their axes are aligned along the magnetic field lines. It is important that in the space between the vortices the material remains superconducting. This is where the electric current flows, so the electrical resistance is still absent. At the same time, there is no Meissner effect; that is, the magnetic field penetrates into the bulk of the superconductor in the form of vortex lines.

Inside a vortex, the magnetic field is nonzero, so every vortex line has a magnetic flux $H\pi a^2$ associated with it. The picture, as we have mentioned, is quite analogous to the flow of current in a superconducting ring. Note that the flux through the vortex core is the same for all vortices and equals twice the magnetic field quantum Φ_0. Thus, every vortex carries one flux quantum, and it is in the form of these units that the magnetic field penetrates into a superconductor.

As the field increases ($H > H_{c1}$), the volume taken up by the normal regions increases, the vortex lines get closer together, and at some field strength H_{c2} (called the upper critical field) superconductivity is completely destroyed and the material becomes normal.

This distinctive picture of field penetration into Type II superconductors makes it possible to use them to create very strong magnetic fields.

HARD SUPERCONDUCTORS. CRITICAL CURRENT

Superconductivity can be destroyed not only by a magnetic field, but by an electric current as well. Type II superconductors described above can exist in the mixed state and are capable of withstanding strong magnetic fields. However, if we taken an ordinary Type II superconductor (a pure metal doped with point like impurities), its critical current will be zero. That is, the flow of an arbitrarily weak current will be accompanied by heating.

Why is this so? Let us place a Type II superconductor in a strong external magnetic field. It will make a transition into the mixed state and a system of vortex lines will appear. Now if we pass a current (called a transport current) perpendicular to the vortex cores, they will begin to move. Indeed, the magnetic field which has penetrated into the superconductor and now resides in the vortices will act on the passing current with the Lorentz force. The current will act back on the vortices with the same force, and the latter will start moving. This leads to dissipative losses in the superconductor, and a voltage drop is observed. There are several dissipation mechanisms. The main one is connected with the normal phase (the vortex region) moving through the crystal. The normal-phase electrons are scattered by the thermal lattice vibrations, which leads to joule losses. There is also the so-called thermal mechanism of dissipation (Clem, 1968) caused by the fact that vortex motion is always accompanied by energy absorption in the region of the forward boundary of the vortex (the superconducting phase changes into the normal phase). This leads to the appearance of microscopic thermal gradients accompanied by heat flow and energy dissipation.

Thus, the transport of electric current through a superconductor in the mixed state is accompanied by generation of heat, which is equivalent to saying that the critical current is vanishingly small.

The picture changes drastically if the sample contains large inhomogeneities (created, for example, by plastic deformation or by admixture of another phase). Such materials (termed "hard superconductors" or "Type III superconductors") possess large critical currents.

Why does the presence of large inhomogeneities affect the superconducting properties so radically? The reason is that the vortex lines are attracted to the inhomogeneities and get stuck on them. This phenomenon is called "pinning." When transport current flows through a hard superconductor, the vortices will start moving only if the Lorentz force becomes strong enough to overcome pinning and tear a vortex away from an inhomogeneity. It is clear that weak current flow will not be accompanied by vortex motion and the corresponding thermal losses. As the current increases, so does the Lorentz force, until finally (when the critical current is reached) the vortices get ripped off the inhomogeneities and begin to move, generating heat. In this way, massive inhomogeneities (pointlike impurities cannot pin down entities as large as vortices) lead

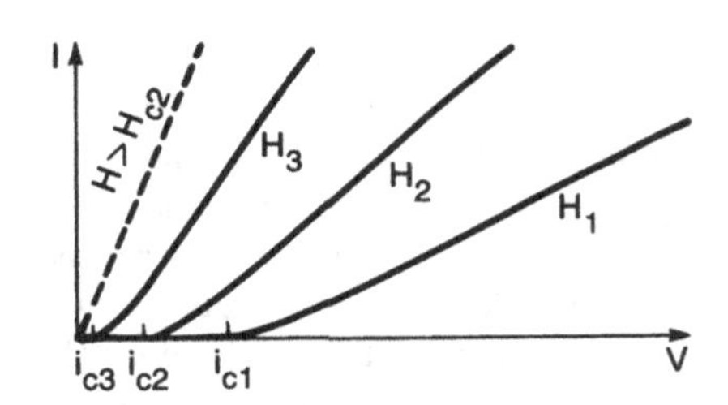

Figure 5.8. Current–voltage characteristics in the presence of the applied field.

to nonvanishing values for the critical current. The properties of hard superconductors are very strongly affected by mechanically working the metals.

The exact magnitude of the critical current depends on such factors as the size of the inhomogeneities and the magnitude of the field. It is possible to obtain large critical currents (see Chapter 12).

We see that owing to the presence of vortices, hard superconductors can withstand large magnetic fields, while structural inhomogeneities make it possible to pass large currents through them.

If the transport current is too large, so that the Lorentz force becomes stronger than the pinning force, the vortices are stripped off, that is, depinned. This activates the energy-dissipating processes described above and results in a finite resistivity (defined as the value of dV/dI and denoted by ρ_f). It should be noted, however, that even though the state is no longer dissipationless as in a usual superconductor, the resistivity ρ_f is still lower than that of the same sample in the normal state, ρ_n. The value of ρ_f depends on the magnetic field and increases with the latter (Fig. 5.8); it becomes equal to ρ_n at $H = H_{c2}$.

The dependence of ρ_f on the magnetic field is simple:

$$\rho_f = \rho_n \frac{H}{H_{c2}} \tag{5.20}$$

This relation can be used to estimate the value of H_{c2}.

Let us also mention that near H_{c2} there are deviations from Eq. (5.20) due to the so-called peak effect. The resistivity reaches its maximum value at some field $H < H_{c2}$. The origin of this effect is still unknown.

SUPERCONDUCTORS IN A VARIABLE ELECTROMAGNETIC FIELD

Above, we described the behavior of a superconductor in a constant, time-independent field. In this case the electric field vanishes, and we have discussed various aspects of the superconductor's behavior in an external magnetic field. We now turn our attention to the case of a variable external field.

This question is relevant both for methods employed in the investigation of the fundamental physics of superconductivity and for various applications. An example of the former is infrared spectroscopy, the technique used for the first experimental observation of the energy gap. Among applications, one can mention transmission lines and microwave cavities. So it is natural that the effect of variable fields on superconductors was one of the first to be considered following the appearance of the BCS theory (Matis and Bardeen, 1958; Abrikosov *et al.*, 1959) (see Bardeen and Schrieffer in Gorter, 1964, also Abrikosov, 1988).

As opposed to the dc case, both the electric and the magnetic components are finite when the external field is variable. (As is known from classical electrodynamics, it is impossible for only one of these components to be finite.) In addition, this field gives rise to certain dissipative processes.

Consider a superconductor placed in a variable field of frequency ω. We need to calculate the current $\mathbf{j}(\mathbf{r}, \omega)$ and the conductivity σ, which turns out to be a complex frequency-dependent quantity. This analysis requires rather elaborate calculations, many of which have to be done numerically. The number of limiting cases also turns out to be large, because the picture depends on a number of parameters: the temperature T, the energy gap Δ, and the frequency (more precisely, the energy $\hbar\omega$). There are a large number of regions corresponding to different relative magnitudes of these parameters. In addition, the behavior depends on the relationship between the penetration depth δ and the pair dimension ξ_0.

As is well known, the behavior of a normal metal in a time-varying field is also quite special. The field cannot penetrate deep inside the metal and is confined to a layer of depth δ called the skin depth and given by $\delta = (c^2/2\pi\omega\sigma)^{1/2}$, where c is the velocity of light and σ is the conductivity. Of particular interest is the situation (realized at high frequencies and low temperatures in pure metals) when the skin depth becomes less than the mean free path. This is called the anomalous skin effect; in this regime, the current depends on the field in a nonlocal way.

In superconductors we end up with a similar situation. What is different is the behavior in the constant-field limit, $\omega \to 0$. In a normal metal, the penetration depth becomes infinite, corresponding to the well-known fact that a dc magnetic field penetrates into the interior of a normal metal. On the other hand, in a superconductor even in a constant field the penetration depth remains finite (the Meissner effect).

The question of electromagnetic absorption in superconductors is most interesting. At $T = 0$ K, absorption takes place only above the threshold frequency $\omega_g = 2\Delta(0)/\hbar$ (see Fig. 5.9). This frequency is in the far-infrared part of the spectrum. At $T = 0$ K, the system has no thermal excitations, and absorption is possible only via the breakup of a Cooper pair. The situation gets more complicated at finite temperatures. First of all, the system now contains thermal excitations (the normal component, in the language of the two-fluid model) which can

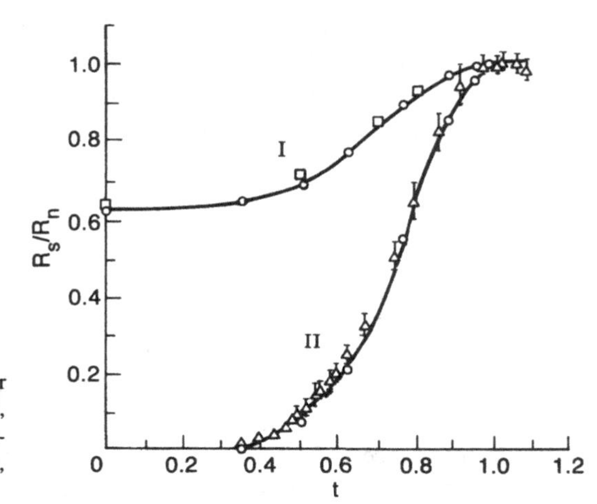

Figure 5.9. Surface resistance for Zn: curve I, $\hbar\omega = 4.3T_c$; curve II, $\hbar\omega = 3T_c$; □ and △, theory; ○, experimental data (Zemon and Borse, 1966).

absorb radiation even at frequencies such that $\hbar\omega$ is less than the pair-breaking energy of $2\Delta(T)$. Nevertheless, if we start increasing the frequency so that finally it exceeds the threshold $\omega_g = 2\Delta(T)/\hbar$, the other absorption channel (pair break-up) will kick in and boost the absorption. It is important that the gap $\Delta(T)$ is temperature dependent and decreases with increasing T (see Chapter 2). As a result, the threshold ω_g can be attained in two ways. The first is to keep increasing the frequency (at fixed temperature) and the second is to raise the temperature (at fixed frequency). Increasing the temperature reduces the gap $\Delta(T)$, so that even if the frequency ω is less than $\Delta(0)/\hbar$ at near-zero temperatures, at some point ω_g will decrease sufficiently for the second absorption channel to turn on. This can be seen in Fig. 5.9.

We see that the presence of an energy gap strongly affects the absorption. Let us mention in this connection that the first experimental verification of the existence of an energy gap was carried out by means of the infrared technique (by Tinkham in 1958) (see Tinkham, 1975).

The behavior of normal as well as superconducting metals is usually analyzed by measuring the so-called surface impedance, defined as the ratio of the electric field at the surface to the integrated current density:

$$\zeta_s = \frac{E(0)}{\int_0^\infty j_y(z)\, dz} \tag{5.21}$$

(the z-axis is perpendicular to the surface). The real part of the surface impedance determines the energy loss of the electromagnetic wave.

The Matis–Bardeen theory analyzes in detail the case which is of the greatest interest for conventional superconductors: when the size of the pairs exceeds the penetration depth (Pippard's case: $\xi_0 \gg \delta$). It should be noted that the small Cooper pair size in the new high-temperature superconductors (so that $\xi_0 \ll \delta$), along with their high anisotropy, requires a different approach (see Chapter 13).

KNIGHT SHIFT

The Knight shift is a well-known and very sensitive method of analyzing internal magnetic fields in solids. Its basic idea is as follows. If we place a metal in an external magnetic field, this field will act on the nuclear spins. As a result, the system will be able to absorb electromagnetic quanta of frequency $\omega \sim \gamma H$, where $\gamma \sim e\hbar(Mc)$ (M is the nuclear mass). This is the phenomenon of nuclear magnetic resonance (NMR). The external field will also act on the conduction electrons. The polarization of the latter will give rise to an additional field acting on the nuclei. As a result, the effective magnetic field seen by the nuclei will be different from the external field, and the resonance frequency will be shifted. This is called the Knight shift (Knight, 1949).

The Knight shift is measured by comparing the nuclear resonance frequency of a nucleus in a metal with the resonance frequency of the same nucleus in a solution, for example, in a paramagnetic salt. In this way, one singles out the contribution of the conduction electrons.

The additional field ΔH is determined by the paramagnetic susceptibility χ and by the probability of finding an electron at the nucleus. More precisely,

$$\Delta H = \left(\frac{8\pi}{3}\right) \chi \, |\psi_S(0)|^2 \, n^{-1} H \tag{5.22}$$

Here n is the number density of the atoms, and $\psi_S(0)$ is the magnitude of the S-wave component of the conduction electrons' wave function at the nucleus. This expression involves only the S-wave component because $\psi(r) \sim r^l$ (l is the angular momentum) and only the $l = 0$ component is nonzero at the nucleus.

When a metal becomes superconducting, the relative change in the Knight shift is proportional to the ratio χ_s/χ_n. It would seem that in superconductors the Knight shift should be vanishingly small at temperatures close to zero. Indeed, the electron system in a superconductor is a set of Cooper pairs, each made up of particles with opposite spins. A weak magnetic field would be unable to polarize this system. However, experimentally a finite Knight shift is observed. This is due to the effect of spin–orbit interaction (Abrikosov and Gor'kov, 1962) (see Abrikosov, 1988), which is observed in small samples (sample size must be much less than the penetration depth, because only in this case can a uniform magnetic field H be set up inside a superconductor). Spin–orbit coupling results in spin no longer being a good quantum number, and the electron system becomes partially polarized.

THE EFFECTS OF STRONG COUPLING. CRITICAL TEMPERATURE

INFLUENCE OF THE PHONON SPECTRUM. ELIASHBERG EQUATION

Many formulas of the BCS theory are of a universal character, for example, $2\Delta = 3.52T_c$ and $\beta = (C_s - C_n)/C_n = 1.43$. In connection with this, we should note first of all that experimentally some pronounced deviations from this universality are observed (see Table 6.1). For example, for lead $\beta = 2.4$, and for the alloy Pb–Bi, $2\Delta = 5T_c$.

The universality of the BCS theory is due to the fact that it was developed in the weak-coupling approximation. In other words, the electron–phonon interaction is assumed to be weak, so the corresponding coupling constant is small: $\lambda \ll 1$. In this approximation, the Hamiltonian, and the entire model, is determined by a single parameter, namely, the quantity λ. For example, the properties of the lattice and the dispersion of phonon curves do not enter directly into the theory. Instead of λ, one can choose another parameter; it is most convenient to choose T_c. Then all the quantities, such as the gap $\Delta(0)$, the heat capacity C_s, and the thermal conductivity κ, will be universally expressible in terms of T_c.

The experimentally observed deviations from this universality are caused by the fact that the electron–phonon coupling is not weak in many materials. It is weak, for example, in Be ($\lambda = 0.2$) and Al ($\lambda = 0.4$). For lead, however, $\lambda = 1.55$, and its properties deviate significantly from the predictions of the usual BCS theory. The same is true for Hg ($\lambda = 1.6$), for the alloys Bi_2Tl ($\lambda = 1.6$) and $Pb_{0.7}Bi_{0.3}$ ($\lambda = 2$), and for many other superconductors. Even in materials with $\lambda < 1$, deviations from the universal BCS relations are observed (for example, for Sn, $\lambda = 0.7$), and a great deal depends on the precision of the measurements.

Table 6.1. Some deviations from the universality of many formulas of the BCS theory.

Material	$\bar{\omega}$ (meV)	$\left(\frac{2\Delta(0)}{T_c}\right)_{\text{1th}}$	$\left(\frac{2\Delta(0)}{T_c}\right)_{\text{1exp}}$
Pb	4.5	4.25	4.3
Hg	1.8	4.7	4.6
In	6	3.66	3.68
Tl	4	3.7	3.6
Sn	6	3.66	3.7
$Pb_{0.88}In_{0.12}$	4	4.34	4.3
$Pb_{0.7}Bi_{0.3}$	4	4.65	4.86

Thus, a theoretical approach is needed which goes beyond the limits of the model with weak electron–phonon interaction. In other words, there is a need for a theory of strong coupling.

The foundation of the theory of strong coupling is formed by an equation derived by Eliashberg in 1960. The Eliashberg equation is complicated, but we shall write it out because of its importance. It has the following form:

$$\Delta(\omega_n)Z = \pi T \sum_{n'} \int d\Omega \, \frac{g(\Omega)}{\Omega} \, D(\Omega_n - \omega_{n'}, \Omega) \, \frac{\Delta(\omega_{n'})}{[\omega_{n'}^2 + \Delta^2(\omega_{n'})]^{1/2}} \tag{6.1}$$

Here $\omega_n = (2n + 1)\pi T, D = \Omega^2[\Omega^2 + (\omega_n - \omega_n')^2]^{-1}$, and Δ is the energy dependent order parameter, which should not be confused with the energy gap. The energy gap can be calculated as the root of the equation $z = \Delta^*(-iz)$, where Δ^* is the analytical continuation of the function of the discrete variable $\Delta(\omega_n)$ onto the upper half-plane; $Z(\omega_n)$ is the so-called renormalization function, which describes the ordinary scattering of electrons by phonons. Equation (6.1) is written for an isotropic system. We shall not write out the equation for Z. The Eliashberg equation is valid if $\Omega \ll E_F$, where $\Omega \sim \Omega_D$.

The Eliashberg equation contains a very important quantity, $g(\Omega)$, which can be written in the form:

$$g(\Omega) = \alpha^2(\Omega) \, F(\Omega) \tag{6.2}$$

Here $F(\Omega)$ is the density of states of lattice vibrations, and $\alpha^2(\Omega)$ describes the interaction between the electrons and the lattice. The function $g(\Omega)$ is a very important characteristic of a material, as it describes both the state of the phonon system and the electron–phonon interaction. There exists a special tunneling spectroscopy method which makes it possible to determine the form of this function. Usually, $g(\Omega)$ contains two peaks (see Fig. 6.1); the lower of the two

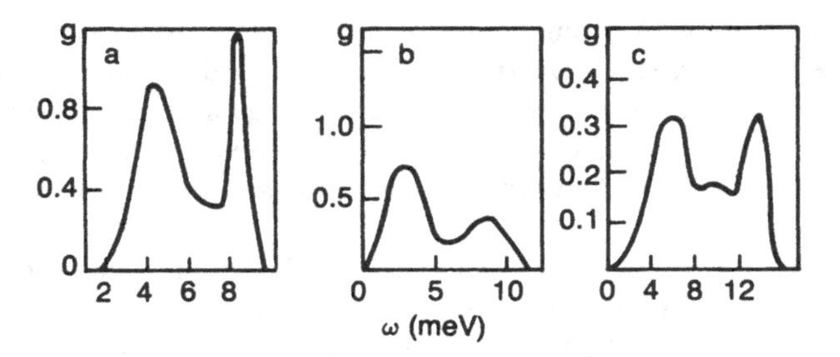

Figure 6.1. The function α^2F for (a) Pb, (b) Bi, and (c) In, obtained experimentally.

corresponds to the transverse phonon branch, and the upper one to the longitudinal branch.

For small values of q, the behavior of phonon branches is purely acoustic: $\omega_t = u_t q$ and $\omega_l = u_l q$. Here $g(\Omega) \sim \Omega^2$. But then for short wavelengths dispersion is observed, and there is a region where the phonon frequency is almost independent of q. In this region, the phonon density of states $q^2\, dq/d\Omega$ has a sharp peak (transformed into a sharp maximum because of damping). The peak regions are most important for superconducting pairing.

The Eliashberg equation is nonlinear (or, more precisely, it is a system of nonlinear equations, if we take into account the equation for Z). We should point out that in the case of weak coupling ($\lambda \ll 1$), corresponding to the condition $T_c \ll \tilde{\Omega}, \Omega \sim \Omega_D$, this equation leads to the results of the BCS theory.

MAIN RELATIONS OF THE THEORY OF STRONG COUPLING

Let us present several relations obtained in the theory of strong coupling. As mentioned above, the primary features of this theory are the loss of the universality characteristic of the BCS model and the appearance of the parameters' dependence on the phonon spectrum. Geilikman and Kresin (1966) obtained the following expression for the ratio $2\Delta(0)/T_c$:

$$\frac{2\Delta(0)}{T_c} = 3.52 \left[1 + \alpha \left(\frac{T_c}{\tilde{\Omega}} \right)^2 \ln \frac{\tilde{\Omega}}{T_c} \right]; \qquad \alpha \simeq 5.3 \tag{6.3}$$

Here $\tilde{\Omega}$ is the characteristic frequency of lattice vibrations. For example, in the presence of two distant peaks, $\tilde{\Omega}$ is equal to the frequency of the lower peak. The second term in brackets is the correction due to strong coupling.

From the relation in the ordinary theory $2\Delta(0) = 3.52T_c$, it follows that the energy gap at $T = 0$ K and the critical temperature are proportional to each other.

In superconductors with strong coupling, as is evident from Eq. (6.3), no such simple dependence exists. If external pressure is applied to a superconductor, this changes both the critical temperature T_c and the quantity $\Delta(0)$. Here one can check whether these values change by the same amount. A study conducted by Zavaritskii *et al.* (1972) demonstrated that $\Delta(0)$ and T_c change in accordance with Eq. (6.3).

Equation (6.3) is not valid for all values of the coupling constant λ; it holds for $\lambda < 2$. What will happen to the ratio $2\Delta(0)/T_c$ if we keep increasing the coupling strength? The ratio will keep increasing, but for $\lambda \gg 1$ it will saturate. This fact can be proven in full generality (Kresin, 1987), and the limiting value of $2\Delta(0)/T_c$ turns out to be about 13.2.

The temperature dependence of the gap also turns out to be entirely different. When the temperature approaches the critical value, the gap decreases in accordance with the law $\Delta(T) = a[1 - (T/T_c)]^{1/2}$. In the usual BCS theory, the value of a is the same for all superconductors and equals $3.06T_c$. For superconductors with strong coupling, this coefficient varies for different superconductors and is given by $a = 3.06[1 + 8.8(T_c^2/\tilde{\Omega}^2)\ln(\tilde{\Omega}/T_c)]$. For example, for lead, $a \simeq 4$.

The large value of the coefficient a explains the substantial deviation of the experimental jump in heat capacity from the BCS model value, $\beta_{BCS} = 1.43$. If, at a given temperature, we compare the entropies S_s and S_n in the superconducting and normal states, respectively, it is clear that $S_n > S_s$. Indeed, in the superconducting state the electrons are in a bound state, their movement is more ordered, and therefore the entropy, as the measure of disorder, is greater in a normal material. The difference $S_n - S_s$ is determined by the size of the gap—the basic parameter that describes the superconducting state. It turns out that $(S_n - S_s) \sim \Delta^2$, and thus $(S_n - S_s) \sim a^2$. Since entropy change is given by $dS = T(dQ/dT)$ (dQ is the quantity of heat), then by using the formula $c = T(dS/dT)$, it is easy to determine β, the jump in heat capacity. It turns out to be proportional to the quantity a^2.

In the usual BCS theory, $a^2 = 9.36$, whereas for lead, for example, $a^2 = 16$. As a result, the jump in heat capacity in superconductors with strong coupling also turns out to be significantly greater than $\beta_{BCS} = 1.43$. A more detailed analysis, based essentially on the considerations presented here (in addition, in the theory of strong coupling, deviations from the usual combinatorial expression for the entropy must be taken into account), leads to the expression

$$\beta = 1.43\left[1 + b\left(\frac{T_c}{\tilde{\Omega}}\right)^2\left(\ln\frac{\tilde{\Omega}}{T_c} + \frac{1}{2}\right)\right]; \qquad b \simeq 18 \tag{6.4}$$

For example, for Pb we obtain $\beta = 2.6$, in accordance with experimental data.

THE FUNCTION $g(\Omega) = \alpha^2(\Omega)F(\Omega)$. TUNNELING SPECTROSCOPY

Let us now consider in more detail the function $g(\Omega)$ which enters the Eliashberg equation. This function (sometimes called the Eliashberg function) plays an exceptionally important role in the theory of superconductivity. This function is a product of two factors: $F(\Omega)$, the density of phonon states, and $\alpha^2(\Omega)$, which describes the coupling between electrons and the phonons of a given frequency.

Suppose we know the form of $g(\Omega)$. What information about the superconductor does it provide us with? First of all, we can determine the coupling constant with the help of the following simple relation:

$$\lambda = 2 \int \frac{d\Omega}{\Omega} g(\Omega) \tag{6.5}$$

In this way, we can immediately determine the strength of the electron–phonon interaction.

Furthermore, the shape of $g(\Omega)$ (see Figs. 6.2–6.4) can tell us which part of the lattice vibrational spectrum makes the principal contribution to superconductivity in a given material. The answer varies from one superconductor to another and can be obtained from an analysis of $g(\Omega)$.

How does one determine the function $g(\Omega)$? There exists a special method developed by McMillan and Rowell (1965) which lets one recover this function based on tunneling spectroscopy data. Over the years, this method has been refined and new techniques have been developed, such as the use of artificial barriers, proximity tunneling spectroscopy, and new methods of data analysis, so that at the present we have at our disposal a new, powerful method of exploring superconductors.

The method is based on the following idea. Consider an S–N tunneling junction (we described the properties of such junctions in Chapter 3; see also Fig. 3.1). Its conductivity σ^S is given by the following expression:

$$\sigma^S = \sigma^N \nu(V) \tag{6.6}$$

where σ^N is the junction conductivity in the normal state, and ν is the so-called tunneling density of states, equal to

$$\nu(E) = \frac{(E)}{[E^2 + \Delta^2(E)]^{1/2}} \tag{6.7}$$

The conductivities σ^S and σ^N can be measured directly. In this way, one can determine the tunneling density of states $\nu(E)$ and then calculate $\Delta(E)$. This is the

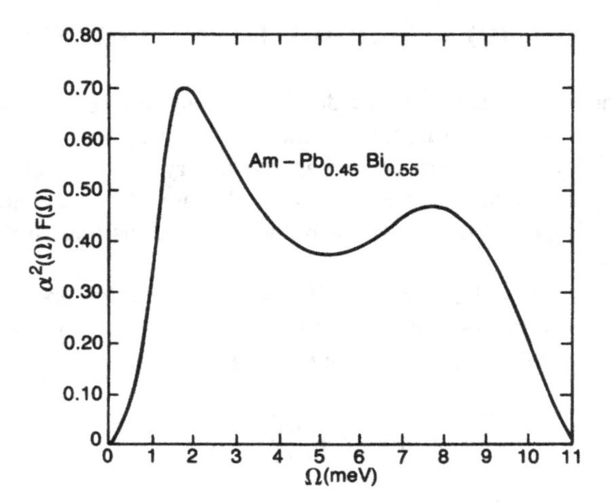

Figure 6.2. The function of $\alpha^2 F$ for Am–$Pb_{0.45}Bi_{0.55}$.

function (more precisely, its analytic continuation) which enters the Eliashberg equation (Eq. 6.1).

Further analysis requires inverting the Eliashberg equation. Initially, one uses a trial function $g(\Omega)$ together with trial values of μ^* and $\Delta(0)$ and calculates (by iteration) the function $\Delta(E)$. It is then used to calculate the conductivity $\sigma(\omega)$.

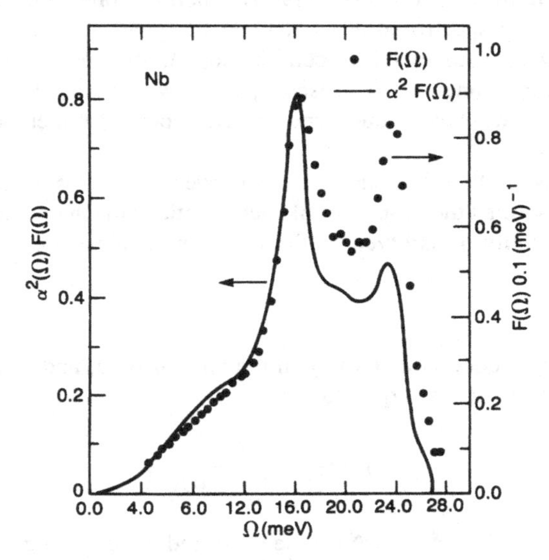

Figure 6.3. Tunneling and neutron spectroscopic data for Nb.

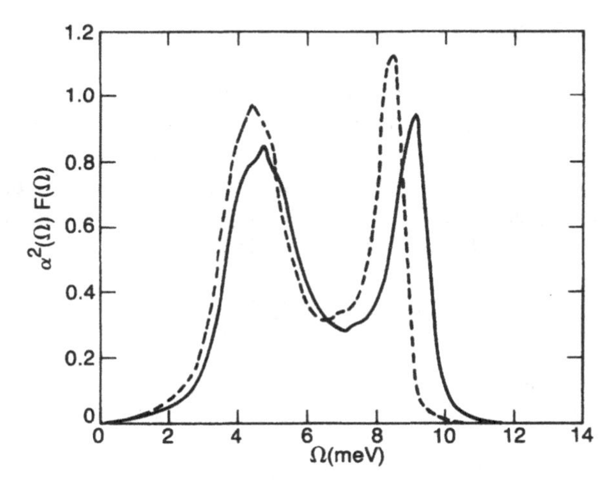

Figure 6.4. Effect of pressure on the phonon spectra of Pb (Svistunov *et al.*, 1981): α^2F for Pb at zero pressure (— — —) and at 12.2 kbar (———).

Comparing the deviation of this calculation from the measured form of $\sigma_{\text{exptl}}(\omega)$, one improves the trial function $g(\Omega)$. The full computer program has been documented by Hubin and is located at the University of Illinois at Urbana.

At present, the function $g(\Omega) = \alpha^2(\Omega)F(\Omega)$ has been determined for many superconductors. Some examples are shown in Figs. 6.2–6.4.

Note that the phonon density of states can be determined independently (by inelastic neutron scattering). Comparing $F(\Omega)$ obtained in this way with $g(\Omega) = \alpha^2(\Omega)F(\Omega)$ allows one to determine the function $\alpha^2(\Omega)$ which describes the electron–phonon coupling. This function is usually smooth relative to $F(\Omega)$.

A comparison of the data obtained by these two completely different methods (neutron spectroscopy, which has nothing to do with superconductivity and provides a direct analysis of the lattice vibrational spectrum, and tunneling spectroscopy, which is a direct inversion of the Eliashberg equation for pairing) reveals in many materials a remarkable agreement of the spectral features. This agreement is powerful testimony to the fact that superconductivity in these materials is indeed due to electron–phonon interaction.

Very often, the function $g(\Omega)$ will contain two peaks. These peaks correspond to the longitudinal $\Omega_{\parallel}(q)$ and transverse $\Omega_{\perp}(q)$ phonons.

The peaks show up at the frequencies where the dispersion $\Omega(q)$ deviates from the usual acoustic law. A vanishing derivative $d\Omega/dq$ corresponds to a delta function-like peak in the density of states. Damping turns this peak into a sharp maximum which shows up in the phonon density of states $F(\Omega)$ and consequently in the function $g(\Omega)$.

CRITICAL TEMPERATURE

We pointed out earlier that the BCS formula (2.6) for the critical temperature holds in the weak-coupling approximation ($\lambda \ll 1$). The Eliashberg equation allows one to examine the dependence of T_c on λ and on the character of the phonon spectrum for any strength of the electron–phonon coupling. An analysis of this equation leads to the following expression for the critical temperature, valid for an arbitrary value of λ (Kresin, 1987):

$$T_c = \frac{0.25\,\tilde{\Omega}}{(e^{2/\lambda_{\text{eff}}} - 1)^{1/2}} \tag{6.8}$$

Here $\tilde{\Omega}$ is the characteristic phonon frequency, equal to $\langle \Omega^2 \rangle^{1/2}$, where $\langle\ \rangle$ implies averaging over the phonon spectrum. Specifically, $\langle \Omega^2 \rangle = (2/\lambda)\int d\Omega\, g(\Omega)\Omega$.

The effective coupling constant is given by

$$\lambda_{\text{eff}} = \frac{\lambda - \mu^*}{1 + 2\mu^* + \lambda\mu^* t(\lambda)} \tag{6.9}$$

The function $t(\lambda)$ is shown in Fig. 6.5; μ^* is the Coulomb pseudopotential (see Chapter 2).

In a simplified model where μ^* is neglected, T_c is given by

$$T_c = \frac{0.25\,\tilde{\Omega}}{(e^{2/\lambda} - 1)^{1/2}} \tag{6.10}$$

We shall now consider some particular cases of Eqs. (6.9) and (6.10), corresponding to different strengths of the electron–phonon interaction, that is, to different values of λ. These special cases had been studied earlier, so it is quite important to consider them in some detail.

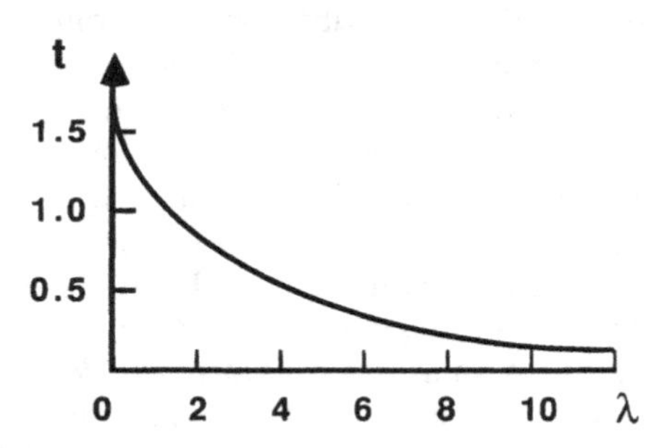

Figure 6.5. Universal function $t(\lambda)$. It can be represented closely by $t(\lambda) = 1.5\exp(-0.28\lambda)$ (S. Tewari and P. Gumber, 1990).

Weak and Intermediate Coupling ($\lambda < 1.5$)

Consider first the case of weak coupling ($\lambda \ll 1$). For simplicity, let $\mu^* = 0$. Then we can neglect the 1 in Eq. (6.10), and we recover the BCS model dependence $T_c \sim e^{-1/\lambda}$ [cf. Eq. (2.6)]. The preexponential factor is different; this is due to the fact that in strong-coupling theory [Eq. (6.1)] one does not assume that the interaction is constant all the way up to Ω_D. It is in this sense that one says that the BCS formula (2.6) is preexponentially accurate.

Note that weak coupling does not at all imply low critical temperatures. T_c is determined by the interplay of two quantities: the coupling strength λ and the energy scale (the extent of the phonon spectrum). One also has to remember that these quantities are not independent [see below, Eqs. (6.16) and (6.17)]. In principle, a large energy scale can provide high T_c even if λ is small.

If $\mu^* \neq 0$ (typically, $\mu^* \approx 0.1$), one has to employ Eq. (6.8). One can neglect the 1 up to $\lambda < 1.5$. Then one obtains

$$T_c = 0.25\, \tilde{\Omega}\, e^{-1/\lambda_{\text{eff}}} \tag{6.11}$$

where λ_{eff} is defined by Eq. (6.9) with $t \approx 1.2$–1.5 (see Fig. 6.5). The values of T_c obtained from Eqs. (6.9) and (6.11) are quite close to those derived from the McMillan–Dynes equation (McMillan, 1968a; Dynes, 1972) (see Parks, 1968):

$$T_c = \frac{\langle\Omega\rangle}{1.2} \exp\left[-\frac{1.04\,(1+\lambda)}{\lambda - \mu^*(1 + 0.62\lambda)}\right] \tag{6.12}$$

(One should not expect complete agreement because the coefficients in the McMillan–Dynes formula were picked so as to achieve the best description of systems with Nb-like spectra; for a detailed discussion, see Kresin, 1987.)

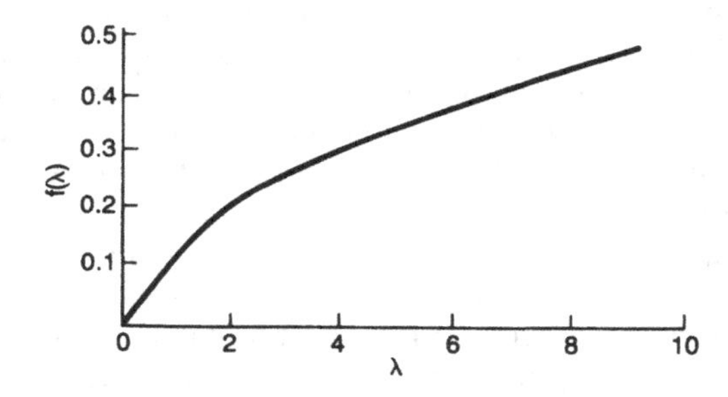

Figure 6.6. Universal function $f(\lambda)$ describing the dependence $T_c(\lambda)$.

Superstrong Coupling ($\lambda \gg 1$)

In this case, one can expand the exponent in the denominator of Eq. (6.8) and obtain

$$T_c = 0.18\lambda_{\text{eff}}^{1/2}\tilde{\Omega} \tag{6.13}$$

where

$$\lambda_{\text{eff}} \cong \frac{\lambda}{1 + 2.6\mu^*} \tag{6.14}$$

If μ^* is neglected, then

$$T_c = 0.18\lambda^{1/2}\tilde{\Omega} \tag{6.15}$$

Equation (6.15) was first obtained by Allen and Dynes (1975). An analytical calculation based directly on the Eliashberg equation also leads to Eq. (6.15) and, in addition, allows the effects of Coulomb repulsion to be taken into account [see Eq. (6.13); Kresin *et al.*, 1984].

It can be seen directly from Eq. (6.13) that, with sufficiently strong coupling, the critical temperature can exceed the Debye temperature. In this case, the dependence of T_c on the coupling constant λ is not exponential [as in Eqs. (2.6) and (6.11)], but assumes the form of a radical: $T_c \sim \lambda^{1/2}$.

One sees from Eqs. (6.9)–(6.14) that the Coulomb term decreases the effective constant ($\lambda_{\text{eff}} < \lambda$), but this decrease differs in a striking way from that in the weak-coupling approximation. In the latter case, the effective constant has the form $\lambda_{\text{eff}} = \lambda - \mu^*$. In the strong-coupling limit, the decrease is not given by a difference, but is described by the ratio given in Eq. (6.14), which gives a stronger dependence on μ^*. If the effect of the Coulomb interaction were described by the difference $\lambda - \mu^*$, as in the case of weak coupling, then this effect would be negligibly small in the limit $\lambda \gg 1$. In fact, an increase of the electron–phonon coupling is accompanied by a transition to the dependence represented by Eq. (6.14), which is more drastic in the limit $\lambda \gg 1$ than simple subtraction. As a result, the Coulomb term makes a noticeable contribution despite the fact that $\lambda \gg 1$. Equation (6.13) holds for $\pi T_c \gg \tilde{\Omega}$, that is, for $\lambda > 5$.

As we pointed out earlier, the ratio $f = 2\Delta(0)/T_c$ increases with the electron–phonon coupling strength, but saturates at $f_{\max} \approx 13$.

A question that arises is what is the maximum value of T_c which can be provided by the phonon mechanism? This question has an interesting history; the answer is not so simple. Up to the early 1960s it was thought that λ could not exceed 0.5 because at higher values of λ the lattice would become unstable. This is incorrect; at present, we know a number of superconductors with values of λ greater than 0.5. These are the main subject of this chapter. On the other hand, if we use the McMillan–Dynes equation [Eq. (6.12)], we would be able to obtain a higher limit T_c^{max}. However, this limit would correspond to $\lambda \approx 2$, and the McMillan–Dynes formula itself is not applicable here.

For greater λ values, one must use the expressions (6.8) and (6.13). It shows that within the framework of the Eliashberg equation T_c can grow without limit with increasing λ. Probably, at some point this growth will be limited by lattice instability, but this question has not been studied in detail.

Increasing the intensity of the electron–lattice interaction leads to a number of peculiarities. For instance, the appearance of bipolarons (formations consisting of two electrons separated by an atomic distance and bound in the potential well formed as a result of strong lattice deformation) seems possible. It should be noted, however, that many interesting and important questions in the theory of superstrong coupling remain to be studied.

COUPLING CONSTANT λ

The coupling constant λ which enters expressions (6.8)–(6.15) for T_c is not a universal constant. It depends on the parameters of each particular system. In the BCS model, $\lambda = \nu_F V$, where ν_F is the density of states at the Fermi level, and V is the effective matrix element of the interaction.

An important result was obtained by McMillan (1968). He pointed out that the coupling constant λ depends to a significant degree on the phonon frequency. It increases with decreasing characteristic frequency $\tilde{\Omega}$ (usually, as $\tilde{\Omega}^{-2}$). According to McMillan, λ can be written as

$$\lambda = \frac{\nu_F \langle V \rangle}{M \langle \Omega^2 \rangle} \tag{6.16}$$

where $\langle \Omega^2 \rangle$ is the average square frequency, $\langle V \rangle$ is the matrix element of the interaction, and M is the ion mass (this expression is written for a monatomic lattice).

*In the case of weak electron–phonon coupling, λ can be written as (Geilikman, 1971; Kresin, 1971):

$$\lambda = \sum_i p_F^{-1} \int_0^{k_1} dq\, q \frac{u_i^2 q^2}{\Omega_i^2(q)} \gamma_i(q) \tag{6.17}$$

The integral is taken over the phonon momenta, $k_1 = \min\{2p_F, q_c\}$, $q_c \simeq q_D$ (p_F and q_D are the Fermi and Debye momenta, respectively), u_i is the speed of sound, γ_i weakly depends on q, and ξ is a constant.*

Thus, whereas it had appeared earlier that in order to increase T_c, the characteristic frequency had to be raised [this increases the preexponential factor in Eq. (6.12)], now it became clear that a more substantial role is played by the dependence of λ on Ω. Therefore, a decrease in the characteristic frequency (also called a "softening" of the phonon spectrum) can increase the superconducting transition temperature.

One can see directly from Eq. (6.17) that λ, and consequently also T_c, is strongly dependent upon the phonon frequency $\Omega(q)$. For low q, this dependence is absent, since $\Omega(q) = uq$. However, for high values of q, which are most important for pairing, the dependence $\Omega(q)$ is no longer linear; in this case, the effect of frequency on T_c becomes strong. The increase in T_c with softening of the phonon spectrum is evident from Eq. (6.17).

The study of the properties of superconductors with strong coupling has shown that the usual Debye approximation is quite a crude model. In describing thermal and kinetic properties of crystals at low temperatures, the Debye model is actually a very good approximation. This is not an accident. The reason is that the lattice heat capacity ($c_{lat} \sim T^3$), or the electrical resistance resulting from the scattering of electrons by lattice vibrations, is related to the thermal motion of the ions. Here the primary role is played by vibrational frequencies satisfying the relation $\hbar\Omega \sim kT$. These correspond to the wavelength $\Lambda \simeq \hbar u/kT$ (since the frequency Ω is equal to u/Λ, where u is the speed of sound). For example, with $T = 1$ K, $\Lambda \simeq 10^{-6}$ cm, that is, approximately 100 crystal lattice spacings! Thus, the thermal vibrations (or "thermal" phonons, as they are called) which are excited at low temperatures have low vibrational frequencies, and correspondingly, large wavelengths. Low-frequency vibrations are in fact sound waves. For them, the dispersion law $\Omega = uq$ is fully valid. Consequently, the Debye model, which assumes that lattice vibrations are sound waves with the ordinary acoustic dispersion law, describes the thermal properties of solids very well.

The situation is quite different in the case of superconductivity. The displacement of ions, which leads to interelectron attraction, is not caused by thermal motion, but by the Coulomb interaction between electrons and ions. It exists at $T = 0$ K as well, at which temperature there is no thermal motion. In this case, not only the low frequencies but the entire vibrational spectrum of the crystal takes part in the interelectron attraction. The strongest attraction is due to the excitation of high-energy vibrational quanta, so that the short-wavelength part of the vibrational spectrum is very important. At high frequencies, the vibrational frequency is not proportional to the wave number; there exists significant dispersion, which must be taken into consideration in analyzing the superconducting state.

Superconductivity results from the exchange not of thermal phonons, but of what are referred to as virtual phonons. Therefore, interelectron attraction exists at $T = 0$ K, at which temperature there are no thermal phonons. The usual mechanism of superconductivity is based on an electron distorting the ion system, that is, emitting a virtual phonon, which is then absorbed by another electron. Phonon dispersion curves have regions where the derivative $d\Omega/dq$ is very small [the points where $\Omega(q)$ vanishes are related to the so-called Van Hove singularities]. These frequencies correspond to the maximum phonon density of states.

ONCE MORE ABOUT THE ISOTOPE EFFECT

We have described earlier (Chapter 1) how the isotope effect played a very important role in the understanding of superconductivity. It provided explicit evidence for the role of the crystal lattice and the electron–phonon interaction in the formation of the superconducting state. Here we would like to discuss this effect in more detail.

The isotope effect reflects the influence of the ionic mass on the transition temperature. To begin with, consider a monatomic lattice. Then the lattice vibrational frequency $\Omega \sim M^{-0.5}$. As for the electron–phonon coupling constant, λ, the McMillan formula (6.16) tells us that it is independent of M, since $\lambda \sim (M\Omega^2)^{-1}$.

If we ignore at first the dependence of T_c on μ^*, the former will be given by Eqs. (6.10) and (6.13). The relationship between T_c and the ion mass is seen to be determined only by the phonon frequency Ω, and therefore $T_c \sim M^{-\alpha}$ with $\alpha = 0.5$. Thus, if we replace the isotope of mass M by one of mass $M^* > M$, the critical temperature should decrease accordingly.

What is the experimental picture? Table 6.2 lists the data for several superconductors. For many, α is indeed close to 0.5, but there are also considerable

Table 6.2. Isotope Effect ($T_c \sim M^{-\alpha}$)

Element	α
Sn	0.46
Mg	0.5
Re	0.4
Ru	0 (±0.05)
Zr	0 (±0.05)
Os	0.21
Mo	0.33

deviations from the canonical value. For example, for osmium, $\alpha = 0.21$, and for molybdenum, $\alpha = 0.33$.

One is also struck by an almost complete absence of the isotope effect in ruthenium and zirconium. In uranium, the isotope effect is negative! In superconducting alloys and compounds, the picture is more diverse. In these, deviations of α from 0.5 are the rule rather than the exception.

We see that the isotope effect is a complex phenomenon, and α is not at all a universal quantity.

What is the origin of these deviations? One possible explanation, which, by the way, was popular in the early years after the appearance of the BCS theory, has to do with non-phonon mechanisms of superconductivity. Clearly, if superconductivity were not due to the lattice, there would be no reason to expect an isotope effect. Ruthenium and zirconium look particularly auspicious in this regard. Alternatively, if the lattice does participate in electron pairing, but there also were a contribution from a non-phonon mechanism, in this case there would be an isotope effect, but with a different value of α.

Such a situation is, of course, possible. However, in reality, it would be wrong to draw conclusions about the presence of a non-phonon mechanism based solely on a deviation of α from 0.5. The fact of the matter is that even if superconductivity is due entirely to phonons, there are still many factors which can affect the strength of the isotope effect.

Again, consider a monatomic lattice. The vibrational frequency, as we have already stated, varies as $M^{-0.5}$. But recall that in the expression for T_c that we used above, we neglected the term μ^* which describes the Coulomb repulsion. What we should really do is use Eqs. (2.8), (6.8), and (6.12), which contain μ^*, the Coulomb pseudopotential. It is important to realize that even though this quantity corresponds to the Coulomb repulsion, it nevertheless depends on the lattice vibrational frequency [see Eq. (2.7)].

Thus, if we consider, for example, the BCS equation (2.8), or the McMillan–Dynes equation (Eq. 6.12), and perform the isotopic substitution $M \rightarrow M^*$, this will change not only the preexponential factor, but the exponent as well. It is easy to see that the effect of changing μ^* on T_c is opposite to that of changing the frequency Ω. If $M^* > M$, the frequency will decrease; at the same time, μ^* will also decrease and, in doing so, will to some extent compensate for the decrease in the preexponential factor. The dependence of μ^* on Ω may seem weak, but μ^* enters Eqs. (2.8) and (6.12) in the exponent and therefore has a significant effect. Indeed, if we use the actual values of E_F, μ^*, and Ω, we will find strong deviations from the value $\alpha = 0.5$ (for instance, for V we find $\alpha \approx 0.2$, in agreement with experiment).

Thus, even for a lattice made up of one kind of ions, it is possible to have large deviations from the simple isotope dependence.

The picture is even more interesting in alloys and compounds. In addition to the μ^* factor which we have just discussed, there is another significant circumstance. These materials contain several different varieties of ions, and the different masses do not simply vibrate independently of each other. There is a complicated relationship between the frequencies of the normal modes and the ionic masses. So if we replace one ion by its isotope, $M_1 \rightarrow M_1^*$, this will not at all result in a simple behavior $T_c(M_1)$. As an illustration, consider the case of a simple cubic two-atom lattice (see Maradudin *et al.*, 1958). Such a structure will give rise to optical vibrations with frequency

$$\Omega^2_{\text{opt}} = \sum_{i=1}^{3} \gamma_i \left(\frac{1}{M_1} + \frac{1}{M_2}\right) - \left[\left(\sum_{i=1}^{3} \gamma_i\right)^2 \left(\frac{1}{M_1} - \frac{1}{M_2}\right)^2 + \frac{4}{M_1 M_2} \left(\sum_{i=1}^{3} \gamma_i \cos q_i d\right) \right]^{1/2}$$

Here M_1 and M_2 are the ionic masses, the γ_i are the force constants, and d is the lattice period. Clearly, if we replace M_1 by M_1^*, the frequency change will not at all be universal: it will depend on the mass of the other ion, M_2, and on the force constants. In addition, bond lengths and, consequently, the elastic constants may be affected by the substitution. *A priori,* we may end up with any Ω_1^*, and consequently with any new critical temperature.

Thus, the variety of manifestations of the isotope effect and deviations of α from the value of 0.5 derived in a simple model can all be understood within the framework of the conventional theory based on the electron–phonon interaction.

We have mentioned the curious behavior of superconducting uranium, which displays a negative isotope effect. An even more exotic behavior is observed in the system Pd–H. When hydrogen is replaced by deuterium, H → D, a negative isotope effect is also observed. Here it is even stronger than in U: upon isotopic substitution H → D, the critical temperature of the solution H/Pd $\approx$ 1 increases from $T_c \approx 9$ K to $T_c \approx 11$ K. The detailed mechanism of this anomaly has not been established; it may be due to a value of μ^* higher than in other materials, or to a change in the elastic constants.

An observation of the isotope effect is a certain indication that the crystal lattice is involved in the formation of the superconducting state. However, the degree of involvement of the electron–phonon mechanism cannot be simply correlated with the magnitude of α and its deviation from the value of 0.5 obtained from a simple model. In order to establish the contribution of different pairing mechanisms, one must employ more detailed analytical methods, the foremost being tunneling spectroscopy. We will come back to this question in Chapter 7, devoted to non-phonon superconductivity.

7

NON-PHONON MECHANISMS OF SUPERCONDUCTIVITY

PAIRING MECHANISMS

The problem of non-phonon mechanisms of superconductivity is intimately connected with that of critical temperature. When we are dealing with, say, the Josephson effect or electrodynamics, the essential facts are the existence of pairing and the presence of an energy gap. All the relevant formulas involve just the gap; the nature of the pairing mechanism is immaterial. However, the situation is different when we are discussing T_c. It is expressed directly in terms of the parameters of the pairing interaction. For example, this can be seen directly from the BCS expression (Eq. 2.6): it gives T_c in terms of the lattice vibrational frequency and the electron–phonon interaction constant.

The usual mechanism of superconductivity involves deformation and polarization of the crystal lattice by the electrons in a metal. The electron–lattice interaction leads to interelectron attraction and superconductivity. The crystal lattice plays the role of the intermediate agency undergoing virtual quantum transitions.

The question arises: would it be possible to find another pairing mechanism, for example, a medium other than a crystal lattice that would give rise to interelectron attraction? This question has direct bearing on the problem of attaining higher critical temperatures.

In 1964, there appeared a paper by W. A. Little which played a fundamental role in further development of the physics of superconductivity. This paper introduced the concept that non-phonon superconductivity can be a way of achieving high T_c. Little proposed a model in which superconductivity with high T_c is obtained with the help of a non-phonon-mediated mechanism of attraction.

We will describe the Little model below, but first we would like to make a brief historical digression. The fact of the matter is that the dream of high-temperature superconductivity existed long before the development of the BCS theory. It had been expected that the future theory would not only explain the phenomenon of superconductivity, but also would show whether it was possible to create high-temperature, maybe even room temperature, superconductors. After the appearance of the BCS theory, the initial answer to this question was negative. It was thought (later, by the late 1960s, it was determined that this opinion was in error) that the electron–phonon coupling constant λ could not exceed the value of $\lambda_{max} = 0.5$; otherwise the lattice would become unstable. If this were the case, the BCS formula shows that the exponential factor would lead to T_c approximately an order of magnitude less than the Debye temperature. This would mean that superconductivity is a low-temperature phenomenon. As we have said (see Chapter 6 on superconductors with strong coupling for more details), this conclusion was erroneous, and it became clear that even with phonons one could have high T_c. However, in the early 1960s this was not thought to be the case.

Little's work was important, not only because he considered non-phonon superconductivity, but also because it revived the dream of high T_c. In addition to reviving a dream, this work also placed the entire concept of high-temperature superconductivity on a serious theoretical footing. The paper by Little can be considered the beginning of the search for high-temperature superconductivity.

What, then, is the Little model? It is shown in Fig. 7.1. Little considered a polymer with a main axis, *AB*, with the conduction electrons moving along this axis. In addition, there are side branches, whose electrons perform oscillatory motion. The main feature of this model is the existence of two groups of electrons. One of the conduction electrons (say, e_1) will, by means of the usual Coulomb forces, induce motion of side-branch electrons. This, in turn, will affect the state of another conduction electron (e_2). In this way, there arises an additional interaction between the electrons e_1 and e_2. The picture is quite similar to that observed in usual superconductors, but now the role of the lattice ions is played by the side electrons. They form the auxiliary system that gives rise to pairing of conduction electrons. In this way, a superconducting state develops in the main axis.

Equation (2.6), which determines the critical temperature of the superconducting transition, contains the characteristic energy $\hbar\tilde{\Omega}$ of vibrational transitions involved in the usual mechanism of superconductivity. The energy of electronic transitions in the side branches of the Little model considerably exceeds $\hbar\tilde{\Omega}$, which leads to higher values of T_c. An estimate of T_c can also be made on the basis of the isotope effect formula $T_c \sim M^{-1/2}$ [see Eq. (1.3)]. In the electronic mechanism, the ion mass will be replaced by the electron mass, giving $T_c \sim m_{el}^{-1/2}$, so that a much higher critical temperature is possible.

Little's work was followed by a flood of papers devoted to different aspects

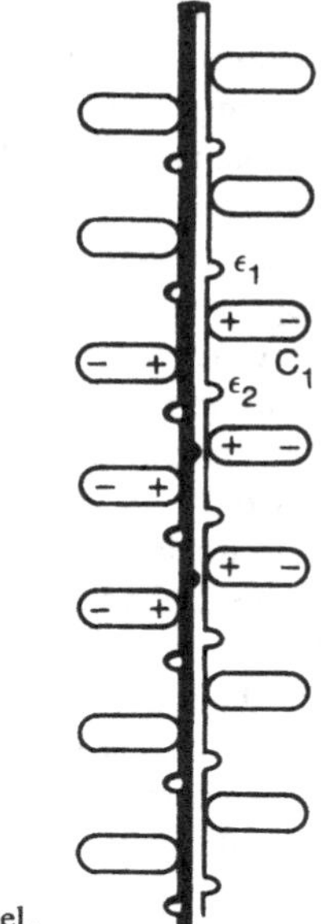

Figure 7.1. Little's model.

of non-phonon superconductivity. It turned out to be important to generalize the one-dimensional model considered by Little to the case of filamentary crystals. In the one-dimensional case involving just one conducting chain, there are strong electron density fluctuations which destroy long-range superconducting correlations. However, if we consider an anisotropic filamentary crystal (we shall call such systems quasi-one-dimensional), then interaction between filaments suppresses the fluctuations; at the same time, all the principal features and advantages of the Little model remain (such as the presence of different groups of electrons, on the conducting axis and on the side branches, leading to pairing).

Ginzburg has proposed a two-dimensional model of a high-temperature superconductor. Specifically, consider a layered system ("sandwich") consisting of a thin conducting film and adjacent nonmetallic layers. Cooper pairs in the metallic film form by exchanging excitations (excitons) in the layers. This model was also considered by Allender, Bray, and Bardeen (1973). It also has been generalized to the case of layered (quasi-two-dimensional) systems, which are favorable for excitonic superconductivity.

Geilikman (1965) showed that the search for high T_c does not necessarily require low-dimensional systems. Under certain conditions, high T_c is possible even in usual three-dimensional systems. He considered two different models.

The first involves superconductors with overlapping energy bands (see Fig. 3.4). Electrons from the A and B zones have different effective masses as well as different wave functions. Both zones have unfilled electronic states, which makes electronic transitions possible. The Coulomb interaction between electrons belonging to different zones gives rise to superconducting pairing. This process can be visualized as follows. An electron in, say, zone A acts with the Coulomb force on an electron in B and excites it; when the B-electron makes a transition into its original state, it changes the state of another A-electron. The net result is the appearance of an additional interelectron interaction in zone A. The system will become superconducting if the resulting attraction is stronger than the usual Coulomb repulsion between the A-electrons. The conditions needed for this mechanism to manifest itself are more restrictive than in the usual case. For this reason, the search for it is a difficult experimental problem. The critical temperature in this model is given by a formula similar to Eq. (2.6), except that instead of the Debye energy, it involves a quantity on the order of the zone width, ΔE_B (more precisely, it is the energy interval corresponding to unfilled levels in zone B). The fact that $\Delta E_B \gg \hbar\Omega_D$ leads to possible high values of T_c.

Another possibility is exemplified by a metal–nonmetal alloy. The conduction electrons can move throughout the crystal, while the electrons in the nonmetallic inclusions form the second system. That is, one is dealing with two groups of electrons, one localized and the other nonlocalized. The localized electrons sit on the impurity centers and are characterized by discrete energy levels. The interaction between these two groups will lead to conduction electron pairing.

Geilikman also proposed the so-called plasmon mechanism of superconductivity (independently put forth also by Fröhlich and Garland). This mechanism is based on the existence of a specific branch of collective excitations which also appears when there are two overlapping energy bands.

Phonons are the best-known example of collective excitations. They describe the vibrations of a crystal lattice. Plasmons also correspond to collective oscillations, but in this case the oscillations are those of the electron system with respect to the lattice. These oscillations are, of course, accompanied by charge fluctuations. Their frequency ω_{pl} in normal metals is very high ($\hbar\omega_{\text{pl}} \approx 5$–$10$ eV) and has the dispersion relation $\omega(q) = \omega_0 + aq^2$. When there are overlapping energy bands, there is an additional plasmon branch, called "demons" [the word "demon" was associated with "distinct electron motion" (DEM) (see Pines, 1956)]. If there are two bands, and two groups of carriers ("heavy" and "light" electrons), then these groups will oscillate with respect to each other. Interestingly, this branch has an acoustic character, so that $\omega(q) \sim q$; that is, there is no gap in the plasmon spectrum at $q = 0$. In this case, the "light" carriers can form pairs by exchanging demons, in full analogy to the usual phonon exchange. The total electron–electron interaction is made up of attraction mediated by the acoustic plasmons and of the usual repulsion, screened by the other "light" electrons.

Therefore, in usual three-dimensional metals there will be acoustic plasmons only in the presence of overlapping energy bands. On the other hand, in low-dimensional systems, such as inversion layers in semiconductors, the plasmon branch does not have a gap even if there is just one energy band. This also leads to the possibility of a non-phonon mechanism in inversion layers (Takada, 1978) (see Ando *et al.*, 1982).

The role of the intermediate system which provides the pairing can be played by a system of spins. The excitations of such a magnetic system are described by quasiparticles called magnons. A magnetic mechanism is probably responsible for the superconducting state in "heavy fermions" (see Chapter 9).

A peculiar situation exists in layered crystals. We will consider this case in more detail in Chapter 13, which is devoted to an analysis of the new high-temperature oxides.

Some curious effects may accompany non-phonon superconductivity caused by the different-electron group mechanism. For example, one may encounter a situation in which one group (say, A) is degenerate (that is, is described by the quantum Fermi statistics) while the other group, B, is described by the Boltzmann statistics. At absolute zero, n_B, the number of B carriers, is zero. As the temperature increases, the energy levels of group B are gradually filling. The additional attraction of group A electrons which is due to their interaction with the B electrons will be very strongly temperature dependent. At absolute zero ($n_B = 0$), there is no attraction. As T increases, the concentration n_B grows, the interelectron attraction in the A group intensifies, and, at some (lower critical) point $T_{c,l}$, it may overcome the repulsive force and superconductivity will set in. It will disappear again at some (upper critical) temperature $T_{c,u}$ when thermal motion will destroy the Cooper pairing of the A electrons. In this way, we may have the curious situation that superconductivity exists in temperature interval $T_{c,l} < T < T_{c,u}$. If such a material is ever found, then at sufficiently high temperatures it will be in the normal state and have a finite resistance. As the temperature is lowered, it will go over into the superconducting state (at $T = T_{c,u}$) and remain in this state down to $T_{c,l}$. At this point the electrical resistance will appear again and remain all the way down to absolute zero. (Actually, near $T = 0$, superconductivity will set in again, caused by the usual mechanism.)

The critical temperature of a superconducting transition is determined by two factors: the energy scale ΔE and the intensity of the interaction λ between the system which becomes superconducting and the system which provides pairing; thus, $T_c \equiv T_c(\Delta E, \lambda)$. According to the usual BCS theory, $\Delta E \approx \hbar\Omega_D$, where Ω_D is the Debye energy and λ is the electron–phonon coupling constant. The basic idea of most non-phonon models has to do with increasing the scale ΔE. In this way, even while keeping λ constant, one can expect a higher than usual value of T_c. When λ is small, one can use the usual BCS expression: $T_c = \Delta E \exp(-1/\lambda)$. If $\Delta E \gg \Omega_D$, one can hope for a high value of T_c.

All this sounds promising, and there is no doubt about the fruitfulness of the concept of non-phonon superconductivity; at first, there were hopes for a quick success story, but reality is much more complicated. For example, increasing the energy scale ΔE affects not only the preexponential factor, but also the magnitude of the coupling constant. This can be seen from the examples of the electron–phonon interaction [see Eq. (6.16)]. Increasing the frequency reduces λ_{el-ph}. An analogous situation arises with non-phonon interactions as well. Of course, the coupling constant λ_{non-ph} depends not only on ΔE, but on other factors as well. For example, the matrix elements describing the interaction between various electron groups are an important factor. It is thus possible to use large ΔE; λ then may be small, but not too small. Therefore, there needs to be a delicate balance between ΔE and λ; this certainly makes it nontrivial to find high-temperature non-phonon superconductivity.

COEXISTENCE OF PHONON AND NON-PHONON MECHANISMS. IDENTIFICATION OF NON-PHONON MECHANISMS

Just a short time ago, the search for high-temperature superconductivity finally paid off with the discovery of a new class of copper oxides with high critical temperatures. The properties of these superconductors will be discussed in detail in Chapter 13.

We would like to point out that it is now understood that high T_c by itself does not imply the presence of a non-phonon mechanism. High critical temperatures can be produced either by the electron–phonon mechanism (in which case the coupling must be sufficiently strong; see Chapter 6) or by a non-phonon mechanism. In addition, the discussion above shows that a non-phonon mechanism does not necessarily lead to high values of T_c. In principle, it is quite realistic for the phonon and non-phonon mechanisms to coexist. Possibly, in some conventional superconductors there is a non-phonon contribution, in addition to the lattice mechanism.

There is, then, a general question that arises in the physics of superconductivity: how does one detect the presence of a non-phonon contribution to pairing and isolate this contribution? This problem is not so simple, because, as we have pointed out, most effects in superconductors are governed simply by the fact that there is a gap, irrespective of how this gap was created.

At first sight, it may seem that one should simply study the isotope effect: $T_c \sim M^{-\alpha}$. However, the isotope effect is a complicated phenomenon; we discussed its peculiarities in Chapter 6. Of course, if the effect is present, that is, if there is a relationship between T_c and the ion mass M, then there is definitely a lattice contribution to the pairing. However, it would be incorrect to try to correlate the value of α (or its deviation from 0.5) with the degree of phonon participation. Even if the phonon mechanism is the only one operative, large deviations from $\alpha = 0.5$ are possible, especially in alloys and compounds. Therefore, the isotope effect does not give an unequivocal answer as to whether a non-phonon mechanism is present.

It is much more informative to employ a method which compares neutron and tunneling spectroscopies. An analysis of inelastic neutron scattering allows the reconstruction of the phonon spectrum, including the phonon density of states $F(\Omega)$. This method directly studies the crystal lattice: neutron scattering has nothing to do with the material being in the superconducting state or not. The function $F(\Omega)$ usually has a number of sharp peaks. These peaks correspond to different branches of the phonon spectrum. For instance, for the longitudinal and transverse acoustic branches, the peaks arise at frequencies corresponding to the dispersion regions where the derivative $d\Omega/dq$ is small, and consequently the density of states, proportional to $dq/d\Omega$, is anomalously high.

Tunneling spectroscopy analysis of the phonon spectrum reflects the effect of lattice vibrations on the superconducting characteristics. This method assumes that only the usual phonon mechanism is present. Tunneling spectroscopy determines the quantity $g(\Omega) = \alpha^2(\Omega)F(\Omega)$, where $F(\Omega)$ is the density of states and $\alpha^2(\Omega)$ describes the electron–phonon interaction. The function $\alpha^2(\Omega)$ is smooth, so that the peaks of $g(\Omega)$ reflect the peaks of $F(\Omega)$. Of course, the presence of $\alpha^2(\Omega)$ modifies the picture somewhat, but nevertheless if the features, and especially the location of the peaks, of the curves $g(\Omega)$ and $F(\Omega)$ (obtained from tunneling and neutron spectroscopies, respectively) coincide, this means that only the usual phonon mechanism of superconductivity is active. Outstanding features, such as additional peaks in the function $g(\Omega)$ which are absent in $F(\Omega)$, indicate that a non-phonon mechanism is present. Indeed, tunneling spectroscopy gives an indirect reconstruction of the spectrum, and if there is a non-phonon mode which lies within the energy interval accessible by tunneling spectroscopy ($\sim$0.1 eV), it will show up as an additional peak. This corresponds to the inclusion of this mode in the Eliashberg equation. For example, such an analysis can detect the contribution of plasmon modes.

Let us consider another example also based on tunneling spectroscopy. Doping a superconductor with complex molecules or adsorbing them onto the surface of superconducting films can lead to additional interelectron attraction accompanied by a rise in T_c. This additional attraction is due to the fact that the molecules have their own internal degrees of freedom. These are, first of all, the vibrational energy levels (in complex molecules, the spacing of these levels is $\sim$100 K). The usual superconducting mechanism involves electrons interacting with the vibrating lattice ions. When the complex molecules are brought in, the intramolecular vibrations will make their own contribution.

It has been experimentally observed (Meunier *et al.*, 1968) that adding complex organic molecules raises T_c. This may be related to the additional interaction described above. In this case, the effect is not large. The main point, however, is as follows. The mechanism of superconductivity based on intramolecular virtual excitations can be detected by the tunneling technique and will manifest itself as an additional peak. The position of the peak can be obtained from the second derivative of the tunneling characteristic. It is important that the molecular frequencies are known independently from molecular spectroscopy. If the position of the peak coincides with the molecular frequency, a new mechanism of superconductivity will be manifested: the effect of intramolecular degrees of freedom on pairing.

SUPERCONDUCTING FILMS

In this chapter we give a brief description of the properties of thin superconducting films. These films play a very important role in applications of superconductivity, especially in superconducting electronics. This will be addressed in more detail in Chapter 12, devoted to applications. In addition, thin films exhibit a lot of interesting physical properties. To begin with, some materials (e.g., Be) are superconductors only if they are prepared in the form of a thin film. Usually, T_c of a film is different from that of the bulk material. For example, bulk Al has $T_c = 1.2$ K, while in films T_c reaches 2.1 K. This is due to the fact that both the phonon spectrum and the electronic characteristics (such as the density of states) in films are different from their counterparts in bulk matter.

Thin films allow one to study the effects of coating. For example, germanium-coated gold becomes a superconductor (Deutcher and Dwir, 1987; see in Wolf and Kresin, 1987). In addition, if the film thickness is less than the penetration depth, one can study a superconductor in a uniform magnetic field.

The properties of thin films often have to be analyzed from the point of view of the physics of two-dimensional systems. The dimensionality depends on the problem under study. If the ratio of film thickness to interatomic spacing is important, then the system is three-dimensional, albeit bound in one direction. However, often what is relevant is the ratio of film thickness to coherence length. In this case, thin films behave as two-dimensional systems.

TWO-DIMENSIONAL SUPERCONDUCTIVITY

The properties of very thin superconducting films are quite different from those of bulk superconductors if the thickness of the film is comparable to the superconducting coherence length. In this interesting situation, at any finite temperature, even at *zero* field, there are vortices present in the film. The reason for this is that the energy necessary to create a vortex excitation is very small if the length of the vortex is less than the coherence length. Thus, there is always a significant number of vortices that are thermally excited in very thin films. There is always an equal number of vortices with clockwise circulation of current (vortex) and anticlockwise circulation of current (antivortex), so that there is no net field at any significant distance from the film. In the absence of pinning, these vortices will always move in the presence of a current because of the Lorentz force acting on the circulating currents, so that the resistance will not be zero at finite temperature. In this situation, where these topological excitations are present in the film, another very interesting phase transition can occur in addition to the Cooper pairing of electrons. This transition involves the pairing of vortices and antivortices, which becomes complete at a finite temperature, T_{2D}. This transition is called the Kosterlitz–Thouless–Berezinski (see J. Mooij in Goldman and Wolf, 1983) transition after the scientists who predicted its occurrence. This type of transition is also observed in two-dimensional helium films and in two-dimensional solids with screw dislocations. This transition in superconductors has some very characteristic features in the physical properties that have allowed it to be studied. The theory of Halperin and Nelson, which is specifically appropriate to superconductivity, predicted that the two-dimensional transition would occur below the bulk superconducting transition of the parent materials, that the separation between T_{2D} and T_{c0}, the bulk transition temperature, would be proportional to the sheet resistance of the two-dimensional film, and that the resistance would be an exponential function of temperature, approaching zero in the limit of very small currents at T_{2D}. Also, at T_{2D} there would be a sharp increase in the superfluid density (n_s), and this would manifest itself in any property of the superconductor (e.g., kinetic inductance) that is sensitive to n_s. Also, the current–voltage characteristic at this temperature would be a pure power law with an exponent very close to 3. At temperatures above and below T_{2D}, the I–V characteristic would be close to a power law but with larger effective exponents below and smaller exponents above T_{2D}. Finally, the resistance versus magnetic field at T_{2D} would be perfectly linear, whereas it would be concave above and convex below T_{2D}. Thus, experimentally, all three measurements must give the same transition temperature. This two-dimensional phase transition has now been observed in granular films and in low-carrier-concentration superconductors, in which it is easy to get the rather large values of the sheet resistance that are important in separating the two-dimensional transition from the ordinary bulk T_c. The role that flux pinning plays at very low temperatures is not perfectly clear at the present time, but it certainly reduces the dissipation that would occur if the vortices were free to move. Both the vortex pairing transition and pinning have the same effect of reducing the dissipation but in totally different ways. Thus, to study the two-dimensional transition, granular films are particularly nice since the pinning is substantially reduced over homogeneous films.

GRANULAR SUPERCONDUCTIVITY

There are a number of unusual effects that can be observed when superconductors are fabricated with an inhomogeneous or granular microstructure. Most of these granular superconductors are in the form of thin films, but several early studies were on packed grains of Nb or on wire drawn from packed powders. Overall, these materials can be classified in several different ways: by the randomness of the coupling between the strong superconducting islands or grains, by the coherence length relative to the grain size, or by the dimensionality of the overall structure.

If the grains are much smaller than the zero-temperature coherence length, then the order parameter is not spatially varying, but its value depends on some average over the properties of the grains and the intergranular material (provided the scale of the inhomogeneities is uniform over the sample). For example, a thin-film mixture of Al and Al_2O_3 prepared by evaporating Al in oxygen can be in this limit. The superconducting properties overall are similar to those of a "dirty" limit superconducting film except that the transition temperature is actually significantly higher than for pure aluminum, having a broad maximum in the moderately resistive region (over 100 Ω per square). This material is characterized by very large resistivities in the normal state and very small critical current densities. This type of superconductor is ideal for probing the metal–insulator and superconductor–insulator transition since the relative amounts of metal and insulator can be varied quite widely.

This material also behaves in an unusual way in a magnetic field. The vortices that are present are typically pinned less strongly than in a homogeneous film. In fact, the pinning force is a nonmonotonic function of the resistivity, having a broad minimum in the moderately resistive regime (the same region where the transition temperature is high). This weak pinning behavior can be useful for devices where the efficient operation of the device requires very free flux motion. A device consisting of two magnetically coupled but electrically isolated granular films can be made into a "dc" transformer. This works on the principle that there is a Lorentz force on vortices in the presence of a transport current. Thus, by passing a dc current in film A, the vortices will move. The vortices in film B will also move. Moving vortices always produce a voltage and a current, so that there will be an induced dc current in film B just by passing a current in film A.

In the situation where the individual grains are much larger than the zero-temperature coherence length, over much of the temperature range the material behaves like a coupled array of strongly superconducting particles, each with a well-defined order parameter, but with the phase from grain to grain dependent on the nature of the intergranular microstructure. Thus, the behavior of the system as a whole depends critically on the manner in which the grains couple together (phase order) to become totally superconducting at lower temperature.

If the coupling between the grains varies randomly from very weak to very strong, then the resistive transition is percolative. As the temperature is lowered, the number of grains that are phase ordered increases as more and more links become superconducting. Thus, clusters of phase-ordered material form and grow so that the length scale describing the average size cluster diverges when a cluster spans the geometrical length of the sample. In simple conductivity percolation theory, the percolation threshold occurs when (as a function of concentration) an infinite cluster of conducting material is formed. In this model, the conductivity varies as

$$\sigma = \sigma(0)(p - p_c)^t$$

where p is the concentration of conducting material, p_c is the critical concentration for the formation of the infinite cluster—the percolation threshold (~17%, 3D; ~50%, 2D)—and t is the power (1.7, 3D; 1, 2D). If we translate this simple model to superconducting bonds during the superconducting transition and make the assumption that the number of superconducting bonds is proportional to temperature, then the resistivity should vary as

$$\rho = \rho(0)(T - T_c)^t$$

Once the infinite cluster is formed, the volume of coupled grains continues to grow. This can be observed by a power law growth of the critical current density below T_c ($R = 0$).

Several granular systems, for example, Al–Ge and Pb–Ge films, can be very well described by this simple model.

There are other granular systems in which the coupling between the grains is not random but varies only over a narrow range of coupling strengths. The resistive transition of such a system is not easily calculated, but the temperature at which the resistance goes to zero relative to the T_c of the grains depends on the strength of the coupling. These latter granular superconductors can, if the resistivity is high enough, undergo a two-dimensional Kosterlitz–Thouless–Berezinski transition involving the pairing of vortex–antivortex excitations (see previous section in this chapter).

These latter materials, consisting of reasonably uniformly coupled grains, behave at low temperatures (temperatures well below the $R = 0$ temperature) like an array of Josephson junctions. They exhibit many of the phenomena that are exhibited by single junctions. They can be used as the "weak" element in a SQUID, they can be quite sensitive to electromagnetic radiation, and they can be readily switched from the superconducting to the normal state (see Chapter 12). There are several examples of granular systems that behave in this manner, but perhaps the most ideal are either a cermet consisting of NbN superconducting grains surrounded by a BN insulating matrix or a similar In–InO mixture.

LOCALIZATION

When metal become strongly disordered so that the elastic mean free path is much smaller than the inelastic mean free path, there are a number of very unusual transport anomalies that occur, and these are often lumped together in the category of localization effects introduced by Anderson in 1958. The simplest picture of what localization means is to think of a very long wire whose resistance is just a function of length. It can be shown that the largest wave packet consisting of fully coherent electrons occupies a length of wire whose resistance is approximately 30,000 Ω. Thus, if the wire contains many such 30,000-Ω segments, then the electrons cannot freely move from one segment to another, but rather they must thermally hop over a "barrier" between segments. Thus, at absolute zero such a "long" wire will be an insulator, even though it is made from a metal. The concept of localization is thus that even in a metal the electrons are localized to wave packets of finite size, the size depending on the disorder. The more disorder, the smaller is the wave packet. These "localization" effects manifest themselves in two and three dimensions in addition to the one-dimensional effects that were just described.

Localization also affects superconducting properties, and the most direct manifestation is in the depression of the transition temperature with sheet resistance in disordered thin films. It has been experimentally shown that superconductivity disappears when the sheet resistance of a film is larger than some value (approximately 10^4 Ω per square). There is still the open question as to whether there can be an insulator–superconducting transition, that is, from a localized state to a superconducting state (see Goldman and Wolf, 1983).

PROXIMITY EFFECT. INDUCED SUPERCONDUCTIVITY

So far in this book we have been talking about what may be called "intrinsic" superconductivity. That is, an isolated specimen makes a transition into the superconducting state as its temperature is lowered below T_c. In addition to this intrinsic kind, there exists also the phenomenon of induced superconductivity. This means that in a material which by itself is not a superconductor one can, under certain conditions, induce the superconducting state. Such a transition occurs owing to the proximity effect which is the subject of this section.

Let us consider a sandwich (Fig. 8.1) consisting of two films, one a superconductor and the other a normal metal (an S–N sandwich). As an example, we

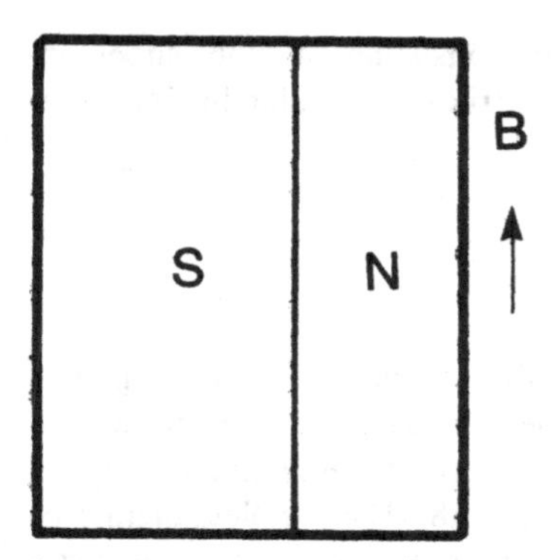

Figure 8.1. *S–N* proximity system.

can take a Nb–Cu contact. By itself, copper is not a superconductor, but as part of the *S–N* sandwich, it begins to display superconducting properties. This happens under the influence of the neighboring *S* film (in our example, niobium). This induction of superconductivity in a normal film is the essence of the proximity effect. The transition of copper into the superconducting state can be observed directly: if the *S–N* sandwich is placed into a magnetic field (Fig. 8.1), copper will exhibit the Meissner effect.

The proximity effect was discovered in 1961, soon after the appearance of the BCS theory.

What is the origin of the effect? Because of the contact, Cooper pairs from the *S* film can penetrate into the *N* film, creating superconductivity in the latter. In other words, the pair wave function Δ (**r**) (this function is called the order parameter and should not be confused with the energy gap) becomes nonvanishing in the *N* film as well.

The proximity effect is interesting from two points of view. First of all, it allows superconductivity to be induced, in materials that normally are found only in the normal state. In addition, the *S–N* proximity sandwich is a model of a spatially nonuniform system, allowing one to study, for instance, the properties of materials with complicated density or charge distributions. The sandwich is nonuniform not only in that it consists of two different materials, but also in that the order parameter $\Delta(\mathbf{r})$ inside each film strongly depends on the coordinates.

It is clear that the degree of the proximity effect depends on the quality of the contact and on the thickness of the films. Since the effect is caused by Cooper pairs penetrating from the *S* film into the *N* film, it is clear that increasing the thickness of the *N* film will weaken the effect. Another important factor is the temperature. Decreasing it diminishes chaotic thermal motion, which in turn facilitates pair penetration. One can introduce an effective coherence length, ξ_N, in the *N* film (Clarke, 1969), such that $\xi_N = hv_F{}^N/T$ ($v_F{}^N$ is the Fermi velocity in the *N* film). The proximity effect is strongest at temperatures far from T_c and close to 0 K. The copper film in the contact NbTi–Cu displays superconductivity up to the thickness $L_{Cu} = 2 \times 10^5$ Å (Mota *et al.*, 1982; see also Mota *et al.*, 1989).

A number of methods have been developed in the theory of the proximity effect. An important contribution has been made by P. de Gennes and his collaborators (the "Orsay group"). Their analysis is based on the Ginzburg–Landau theory. The spatial variation of the order parameter has been taken into account directly, by introducing a special boundary condition at the boundary between the two films.

Another approach, the so-called tunneling model, has been developed by McMillan (1968). According to this model, the proximity effect is generated by electrons tunneling through the barrier formed by the interface region. Interface quality affects the magnitude of the so-called tunneling matrix element. (To describe the tunneling process, one introduces the quantities Γ^S and Γ^N, such that $\Gamma^\alpha = |T|^2\nu_\alpha$, with $\alpha = S$ or N. Here, T is the tunneling matrix element, and ν_S and ν_N are the densities of states in the S and N films, respectively). As opposed to the Orsay group method, which is applicable near T_c, the McMillan method is valid for all temperatures. It was originally developed under the assumption that the order parameters Δ_S and Δ_N are constant, that is, that the film thicknesses are small ($L_S \ll \xi_S$, $L_N \ll \xi_N$). However, the tunneling model can be generalized to greater thicknesses.

As we have already stated, the proximity effect involves a Cooper pair making a transition from the superconducting into the normal film. This transition can be considered as a reflection off the S–N boundary, with an electron incident on the boundary and a hole reflected (Andreev's reflection; 1964). Such a reflection is, of course, equivalent to an electron pair vanishing from the S film (and going into the N film). Indeed, the disappearance of an electron is equivalent to the creation of a hole. Therefore, when describing the disappearance of an electron pair at the boundary, we can term the disappearance of one of the electrons "the creation of a hole" and picture the entire process as the disappearance of one electron accompanied by the creation of one hole. This leads us to the picture of Andreev's reflection, which turns out to be convenient for describing processes taking place at the S–N interface.

The proximity effect is important not only for S–N sandwiches in which the N film is normal (in the isolated state) at any temperature (e.g., Cu, Sb). It shows up also in the analysis of S_α–S_β systems consisting of two different superconductors ($T_c^\alpha \neq T_c^\beta$; for concreteness, let us assume $T_c^\alpha > T_c^\beta$, where T_c^α and T_c^β are the critical temperatures of isolated α and β films). An example of such a system would be the Pb–Al sandwich. An S–N system can be considered a special case of S_α–S_β with $T_c^\beta = T_c^N = 0$ K. Magnetic screening in various proximity systems has been studied by Simon and Chaikin (1981; 1984).

A proximity system is characterized by a single critical temperature T_c, with

$T_c^{\beta} < T_c < T_c^{\alpha}$. The critical temperature is determined by the quality of the contact as well as by such film parameters as their densities of states and thicknesses. For example, for the S–N system ($L_N \ll \xi_N$; $L_S \ll \xi_S$), T_c is given by

$$T_c = T_c^s \left(\frac{T_c^s}{\Gamma} \right)^{\nu_N L_N / \nu_s L_s} \tag{8.1}$$

This formula assumes that $\Gamma > T_c^S$; the tunneling parameter depends on the contact quality.

We have pointed out earlier that the importance of the proximity effect lies in the possibility of inducing superconductivity in materials that are not superconductors by themselves. In doing so, one can utilize both the intrinsic properties of these materials and superconductivity. A particularly interesting case arises when N is a semiconductor [a system of this type, namely, Pb–Te–Pb, was first studied by Seto and Van Duzer in 1972) (see Van Duzer and Turner, 1981). It is very appealing to combine semiconducting and superconducting properties. For instance, by varying the carrier concentration in the semiconductor, one can significantly affect the critical temperature. This can be seen directly from Eq. (8.1). The density of states ν_N depends on the carrier concentration: $\nu_N \sim p_F^N \sim n_N^{1/3}$ (p_F^N is the Fermi momentum in the N film). The carrier concentration can be affected either by radiation or by employing the field effect. The effect of radiation on the properties of superconductor–semiconductor junctions was studied by Barone *et al.* (1974).

The Josephson system S–N–S where N is a normal metal or a semiconductor (e.g., Nb–InAs–Nb) are also essentially proximity systems which are capable of sustaining nondissipative Josephson current. The presence of the N film's own electron subsystem plays an important role. In addition, the N film can be up to several thousand angstroms thick. All this makes S–N–S contacts different from the usual S–I–S Josephson contacts, where I is a thin insulating layer. The proximity effect induces a special superconducting state in the metal (or semiconducting) N film, which makes it possible for weak superconductivity to arise.

The system Nb–InAs–Nb turns out to be especially interesting (Takayanagi and Kawakami, 1985). It was used to observe the field effect (Fig. 8.2). Applying a voltage to InAs changes the magnitude of the Josephson current. The origin of this effect (Kresin, 1986) is the fact that InAs has a native inversion layer. This two-dimensional layer (we are again dealing with a 2D problem) turns out, owing to the proximity effect, to be the channel through which the Josephson current flows. Applied voltage changes the number of carriers in the inversion layer, which exponentially [$\exp(-aN^{-1/2})$] affects the amplitude of the current. This scheme is promising for the creation of a superconducting three-terminal device.

The system S_α–N_β–I–S (for instance, Pb–Cu–PbO–Pb) is again a usual proximity Josephson contact. The Josephson effect arises between the Pb and Cu

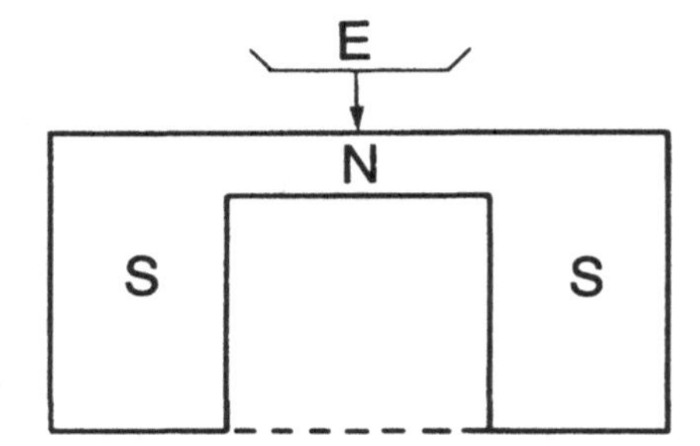

Figure 8.2. Field effect; N is the inversion layer.

films; the latter becomes superconducting owing to the proximity effect. The magnitude of the current and its temperature dependence are essentially defined by the thickness of the normal film.

The proximity effect can also display itself by altering T_c of superconducting films. For instance, it is known that, with decreasing thickness, the T_c of Nb films drops (Wolf *et al.*, 1978). When the film thickness is $L = 120$ Å, $T_c = 6.3$ K; if $L = 40$ Å, $T_c = 4.35$ K. This effect is caused by the presence of the oxide NbO on the film surface. The oxide is a superconductor with $T_c = 1.4$ K. Thus, we are dealing with a proximity system S_α–S_β with S_α = Nb and S_β = NbO. Decreasing the Nb film thickness amounts to decreasing L_α. This increases the relative importance of the layer with lower critical temperature (the β layer), and T_c drops.

PREPARATION OF SUPERCONDUCTING FILMS

It is extremely important to be able to prepare thin films of superconducting materials since much of the physics and many of the applications of superconductivity require them. Both the conventional superconductors and the new ceramic oxides have been successfully prepared by a large number of methods. As can be expected, the techniques that have worked for the conventional superconductors have in general worked for the ceramic oxides. This section deals with a description of most of the techniques that have been used to successfully prepare superconducting films. Most of the systems that are described below are commercially available with a large number of features and options that make them suitable as received for preparing excellent-quality superconducting films on a variety of substrate materials.

Evaporation

Evaporation is conceptually the simplest of all the deposition techniques. In practice, however, some of the most sophisticated apparatuses are used to evapo-

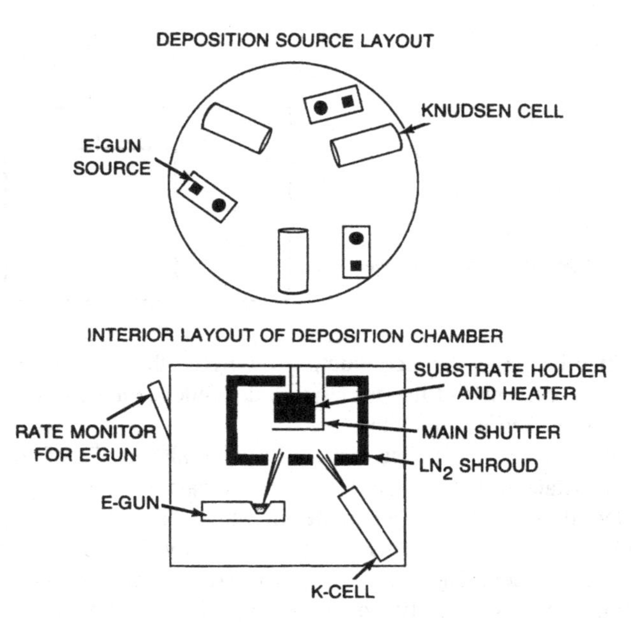

Figure 8.3. A simple schematic view of an evaporation system using both electron beam and thermal sources.

rate epitaxial films† of superconducting materials under very controlled conditions, and these systems are more accurately called molecular/electron beam epitaxy systems (MBE/EBE). The technique involved utilizes a vacuum system to remove most of the contaminating gases from the deposition chamber. Typical pressures that are obtained in simple evaporators are in the 10^{-7} torr range while the MBE/EBE systems require pressures of less than 10^{-10} torr. The elements or compounds to be evaporated are heated in crucibles by either resistive heating elements (tungsten, tantalum, etc.) or by electron beam heating. Typical evaporators have more than one evaporation source, and it is possible to obtain systems with as many as six independent sources. The high temperatures produced in these sources cause the vapor pressure of the evaporant to rise to a level at which a significant amount of these materials can be collected on a substrate that is located on a direct optical path from the evaporant. The substrate can typically be at a variety of temperatures, ranging from 77 K to approximately 1300 K depending on the required microstructure of the final film. In most modern evaporators, the materials to be evaporated from multiple sources can be deposited on the substrate sequentially to make multilayer structures or can be

†Epitaxial films are nearly single crystals that are grown in registry with the atoms in the substrate.

deposited simultaneously to make compound and alloy films from, for example, the A15 structure compounds such as Nb_3Sn or V_3Ga. This technique has also been used quite successfully in preparing films of the high-transition-temperature cuprates, both to deposit the material in the correct stoichiometry for subsequent annealing and to introduce the evaporants in a variation of the traditional evaporation process that has been called reactive evaporation. For reactive evaporation, a small partial pressure of the gas that is to be reacted with the evaporants (oxygen in the case of the cuprate superconductors) is directed toward the substrate by a small tube. This technique has allowed for the complete formation of the superconducting phase of Y–Ba–Cu–O without any further processing. Evaporation was the first method that was successfully used to prepare high-quality films of Y–Ba–Cu–O that exhibited rather high values of the critical current density. The drawbacks of this technique are the difficulties of precisely controlling the correct ratio of constituents in a multicomponent superconductor such as the new cuprates. In order to get a stoichiometric film, the deposition rates of the various evaporants must be precisely controlled, especially in the presence of a reactive gas. Furthermore, there is a definite limit on the partial pressure of oxygen that can be introduced before the oxygen seriously limits the lifetime of the elements used to heat the evaporants and the substrates. A schematic drawing of an evaporation system is shown in Fig. 8.3.

Sputtering

The technique of sputtering again involves the use of a vacuum system to remove the contaminating gases to a very low level (pressures typically less than 10^{-7} torr). The material to be deposited is fabricated into a disk called the target (sometimes the target can be rectangular, or, in some rare cases, it can be cylindrical) and can consist of a single element, an alloy, or a complicated compound such as the new cuprate materials. As in evaporation, a sputtering system can contain many targets, each used sequentially or simultaneously. The targets are mounted on electrically isolated but water-cooled copper plates that are connected to the negative sides (cathodes) of sources of high-voltage (25–2000 V) dc or rectified and impedance-matched rf (13.56 MHz). The system is then backfilled with argon or one of the other inert gases to partial pressures of several millitorr (the limits of the reported values range from 1 to 500 millitorr). When the voltage is applied, a plasma of the inert gas is formed around the target toward which the ions of the gas are accelerated. These ions hit the target with enough energy to knock off atoms or molecules, which are then collected on a substrate located very close to the target. The substrate can be at a variety of temperatures just as in evaporation, depending on the required microstructure of the resulting film. This technique has been used to prepare very high quality films of many of the elemental superconductors including Nb and Va. A variation of this technique called reactive sputtering involves the addition of one or more reactive gases to the chamber, either premixed with the argon or introduced separately. Thus, as in the case of reactive evaporation, the target material can be chemically altered either before it is sputtered, in the gaseous state, or as it is deposited on the substrate. Which of the above-mentioned processes occurs depends on a variety of factors including the relative amount of reactive gas, the sputtering rates, and the substrate temperature. This technique is the preferred method for producing films of NbN and NbCN and has been successfully utilized to deposit high-quality films of the various cuprate superconductors. Sputtering has been very successful in fabricating multiple-layer superconducting structures for making tunnel junctions (see Chapter 12). Most of the superconducting circuits have been made by sputter deposition and patterning (see next section). This technique has been very successful for the conventional low-transition-temperature superconductors but has had some problems when applied to the cuprates. In particular, it has been difficult to control the stoichiometry of the material on the substrate due to the presence of oxygen ions in the plasma. This oxygen plasma changes the sticking coefficient of the various elements. (The sticking coefficient is the probability that an atom striking a surface will remain on the surface.) This problem has been overcome by either adjusting the composition of the target or by a combination of high sputtering pressures and a perpendicular geometry between the target and the substrate. A schematic diagram of the sputtering process is shown in Fig. 8.4.

Ion Beam Sputtering

Ion beam sputtering is closely related to the sputtering technique described above. The major difference is that the plasma is generated in a gun rather than

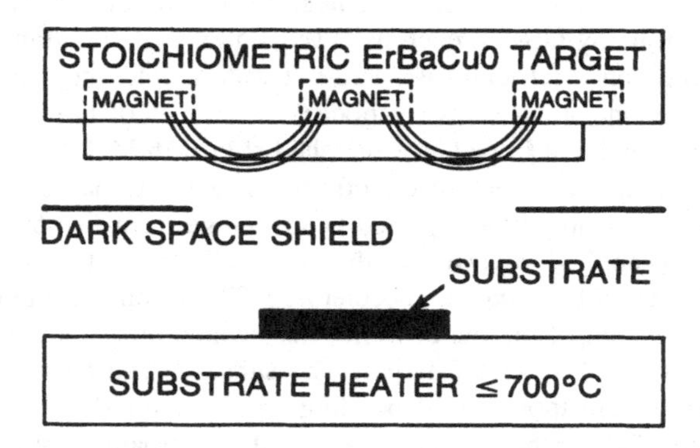

Figure 8.4. A simple schematic of a magnetron sputtering system.

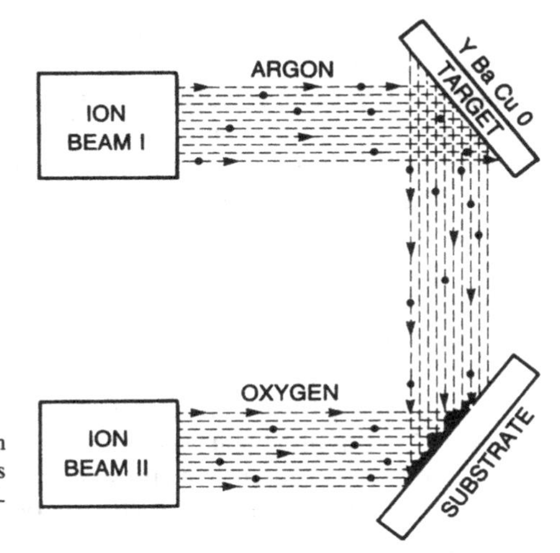

Figure 8.5. A schematic of a dual-beam ion sputtering system. One beam sputters the target; the second beam chemically alters the depositing film.

around the target. In the sputter gun, the argon ions are accelerated by a series of high-voltage grids and then neutralized before exiting the gun at high velocity. The atoms are directed toward a target which is sputtered as described above. The material removed from the target is collected on a substrate located at the appropriate location. Again, several guns can be installed in a system, and they can be utilized for different processes. For example, one gun can be directed toward a target of material to be deposited while a second beam of reactive atoms can be directed toward the substrate (oxygen in the case of the cuprates). Thus, a reactive ion beam process can be used in which the energy of the reactive species can be controlled. Furthermore, the neutralizing grid can be deactivated so that an ionized beam can be used instead of a neutral one. The main advantages of this technique are that the plasma is isolated and does not interact with the substrate and that the beam can also be used to etch patterns in films when unprotected areas of a film are exposed to the beam. This technique has not been widely used so that it is not clear how well it compares to either evaporation or conventional sputtering. A schematic diagram of an ion beam sputtering system is shown in Fig. 8.5.

Laser Evaporation

Laser evaporation is a rather new deposition technique that was being developed for depositing semiconducting layers before the discovery of the cuprate superconductors. This technique is quite simple. The main part of such a system is the deposition chamber, which contains a target consisting of the material that is ultimately going to be deposited. The target can be rotated so that the area that is hit by the laser can be continuously changed. Directly opposite the target is the substrate, mounted in a holder which can be heated to about 1000 K. The source, generally a UV excimer pulsed fluoride laser, is mounted outside a UV-transparent window of the chamber. The laser is focused on the target. Again the chamber is evacuated to fairly low pressures (about 10^{-6} torr) and then backfilled with a reactive gas, if required (for the cuprates, oxygen is introduced into the chamber to pressures in the torr range). The laser energy ablates the target, removing atoms and molecules. Typical laser energies are in the range of hundreds of millijoules per pulse, and the pulse rate can be as high as 20 per second. This technique has produced excellent-quality films of Y–Ba–Cu–O on various substrates heated to 1000 K. These films required no additional processing and have both high critical current densities and very low surface resistance (see Chapter 12). The disadvantages of this technique are the very directional plume of stoichiometric material, which makes coverage of large areas difficult, and the presence of large particulates or clusters on the surface of the substrate that are produced by excessive laser power or poorly prepared targets. A schematic view of a laser evaporation system is shown in Fig. 8.6.

Other Techniques

In addition to the techniques that have been described above, which all fall in the category of physical vapor deposition, there are a number of techniques that are generally described by the term chemical vapor deposition (CVD) and metallo-organic chemical vapor deposition (MOCVD). These techniques involve the preparation of precursor compounds that have high vapor pressures at relatively low temperatures as well as decomposition temperatures that are also fairly low (500–700 K). The apparatus is often made of quartz or Pyrex and consists of several containers for the precursors, each with its own heating element. The

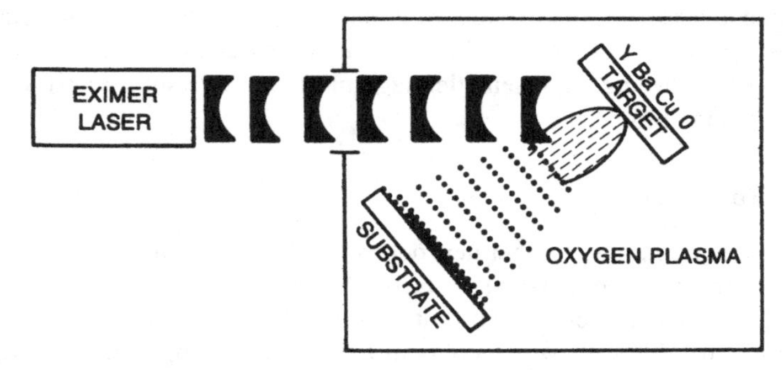

Figure 8.6. A simple schematic of a laser evaporation system.

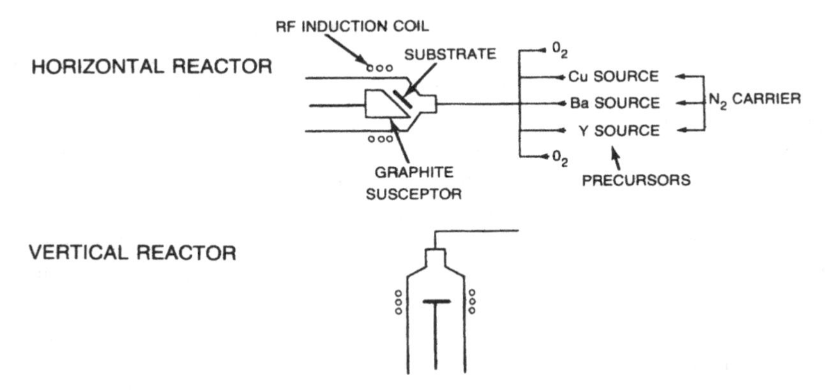

Figure 8.7. A simple pictorial representation of an MOCVD system.

main chamber is called the reactor and consists of a heated platform (normally of graphite), called a susceptor, on which the substrate is placed. The precursor compounds are heated in their separate containers and the vapors are swept into the reactor with a carrier gas, which can either be an inert gas or a gas which can later react with the precursors on the heated susceptor and produce the required material on the substrate. This technique has not been widely used in the preparation of conventional superconductors because the more conventional techniques have proven very successful. Recently, the required precursors have been synthesized for all the elements required for fabrication of the various cuprate superconductors. This technique may provide excellent films *in situ* since the incorporation of oxygen into the film may be quite straightforward. (The incorporation of sufficient oxygen into the growing film is the hardest part of ensuring the *in situ* growth of high-quality films of the cuprate superconductors.) A schematic drawing of an MOCVD apparatus is shown in Fig. 8.7.

Substrates

Implicit in all the descriptions above was the presence of a suitable substrate material on which the superconducting material could be deposited. The conventional metallic superconductors such as Nb and NbN can be deposited on a variety of substrates including sapphire (single-crystal Al_2O_3), MgO, silicon, and GaAs. The most important problems to be considered are that deposition temperatures may destroy the substrate or accelerate any chemical interactions with the substrate. For example, when Nb is deposited on silicon, it must be done at relatively low temperatures to prevent the formation of the silicides. The problem of appropriate substrates for the cuprates has been a much more delicate one. The cuprates are ceramic materials that are extremely brittle and require high-temperature processing to provide both the correct structure as well as the correct oxygen chemistry. Therefore, the substrate must be able to withstand the high processing temperatures without reacting with the forming superconductor, it must have a thermal expansion coefficient that is well matched to the superconductor so that microcracks do not form on cooling, and it must have the required properties that are necessary for the ultimate use of the superconducting film. For example, strontium titanate satisfies the first two criteria very well: it does not react with the cuprates and its thermal expansion coefficient is very closely matched to that of Y–Ba–Cu–O. However, it has a very large dielectric constant and is very lossy at rf frequencies and is thus not suitable for any microwave device applications. Most of the best films obtained to date have been prepared on strontium titanate and, more recently, on lanthanum aluminate, but there may be new substrates developed that satisfy all three criteria.

Patterning

In order to use any of the above-mentioned films for physics experiments or devices, one must be able to pattern them into suitable geometries for measurements or for successive depositions of contacts, protective coatings, insulating layers, and so on. The simplest way to produce a pattern on a film where the dimensional requirements are in the range of 2 μm and above is a technique called photolithography. In this technique, a special photosensitive polymer called photoresist is spun onto the film with the use of a very high speed motor. Typically, this polymeric film is about 1 μm thick. A desired pattern is projected onto this film through a contact mask or through a specially designed microscope. The polymer film is now "developed," and, depending on whether a positive or a negative photoresist has been used, the exposed area either remains or is removed, leaving the desired pattern behind. Thus, the appropriate regions of the superconducting film remain covered with photoresist. The exposed superconductor can now be removed either by a chemical etch or by various plasma or ion beam etches.

SUPERCONDUCTING SYSTEMS

At present, there are over 6000 known superconducting compounds. They can be classified into several groups according to their properties. In this chapter, we shall discuss several such groups. It is, of course, impossible in this book to give a detailed description of superconducting materials or even to cover all their principal classes. We are not going to try to do that. The main thing that we would like to demonstrate to the reader is the great variety of the world of superconductivity. All the materials to be discussed below are superconductors, but they differ widely in their properties, structures, and so on.

A15 MATERIALS

Until the discovery of the cuprate superconductors, most of the highest T_c superconducting materials all had the crystal structure illustrated in Fig. 9.1. The stoichiometry (stoichiometry is defined as the composition mandated by the ideal crystal structure) of this class of materials is A_3B, where A is one of the transition metals such as Nb, V, Ta, or Zr, and the B element comes from the IIIA or IVA column of the periodic table and is a metal or semiconductor such as Sn, Al, Ga, Ge, In, or Si. As can be seen from the figure, the A elements are situated at the corners of a cube and the B elements form three orthogonal chains. The Nb compounds typically have the highest transition temperature, with that of Nb_3Ge at 23 K being the highest one. The T_c of these compounds is quite sensitive to the stoichiometry, and the maximum of the transition temperature corresponds to being right at the 3/1 ratio of the ordered material. For many of these materials, the density of states near the Fermi level is a very sharply peaked function, and since in the BCS theory the T_c is strongly dependent on the density of states at the Fermi level, the transition temperature will depend on just exactly where the

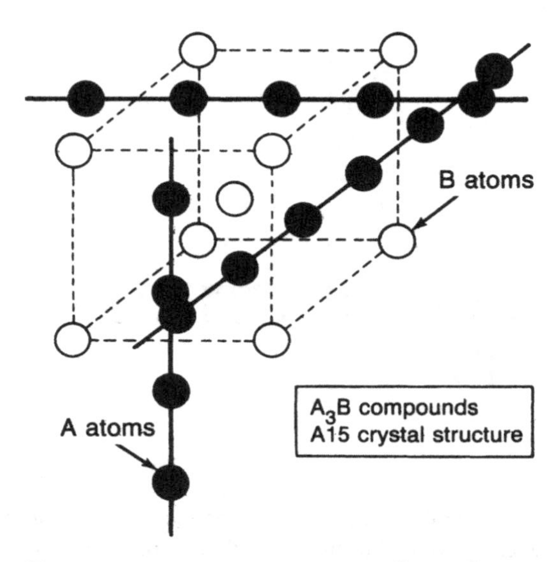

Figure 9.1. Crystal structure of Nb_3Sn: ●, Nb; ○, Sn.

Fermi level is relative to the peak. There also have been speculations that the high T_c is due to the one-dimensional chains giving low-frequency phonons which can also enhance the electron pairing interaction. It is also well known that microstructural disorder produces a smearing of any sharp structure in the density of states, and thus the peaks will be smeared, causing a rapid decline in the transition temperature. This is exactly what is observed for most of these compounds. They are very sensitive to changes in stoichiometry, which not only disturb the location of the Fermi level relative to the peak but also disorder the chains, thus affecting the low-frequency phonons. These materials are also very sensitive to the effects of radiation damage. In fact, the compound with the highest transition temperature, Nb_3Ge, was only found to have the high T_c when it was sputtered. Bulk samples could not be prepared stoichiometrically and had a very low T_c (6 K), but the nearly stoichiometric compound that was formed metastably by high-pressure sputtering had a T_c of 23 K. Interestingly, Nb_3Ge may also be an exotic superconductor because part of its high transition temperature may be due to non-phononic contributions to the pairing (Kihlstrom *et al.*, 1988).

For many years, it was thought that Nb_3Si would have a transition temperature near 30 K. This speculation was mainly based on empirical arguments derived from the tabulated trends for the variation in T_c as a function of atomic size and location of the A and B elements in the periodic table. For example, as the B element is changed from In to Ga to Al (moving up in the periodic table), T_c increases; similarly, moving up in the table from Sn to Ge, the transition temperature increases. Therefore, it was expected that if stoichiometric Nb_3Si could be made, it would have a significantly higher T_c than Nb_3Ge. When Nb_3Si was finally synthesized with close to the 3/1 stoichiometry, the transition temperature was only 20 K. These materials are presently the best for producing very high field superconducting magnets because of the combination of their high T_c and their high critical magnetic field H_{c2} (H_{c2} for Nb_3Ge is 37 T). However, wires of these materials are very difficult to make because of the materials' tremendous brittleness. The problem has been solved for Nb_3Sn (T_c = 18 K) and for V_3Ga (T_c = 16 K) by the use of a technique called the bronze process. Since for stability (see section on power applications in Chapter 12) there needs to be a significant amount of high-thermal-conductivity material surrounding the superconducting filaments, this process, which naturally provides some of this stability, is very attractive. The process is carried out as follows. A Sn or Ga bronze billet (several inches in diameter) is drilled out, and the holes are filled with Nb or V rods (see Fig. 9.2). The ends of this billet are sealed with conical endcaps, and then the billet is repeatedly drawn down into small rods. These rods are then inserted in copper billets that have also been cored. The extrusion or drawing process is continued (with annealing at crucial stages) until a fine wire with many fine filaments of Nb or V is produced. This wire is then reacted to form the A15 compound on the surface of the Nb or V filaments. In some instances, the wire is reacted after it has already been wound into its final configuration, but for magnets with large radii of curvature the wire can be wound after reaction. What is then produced are several thousand filaments of Nb or V each surrounded by

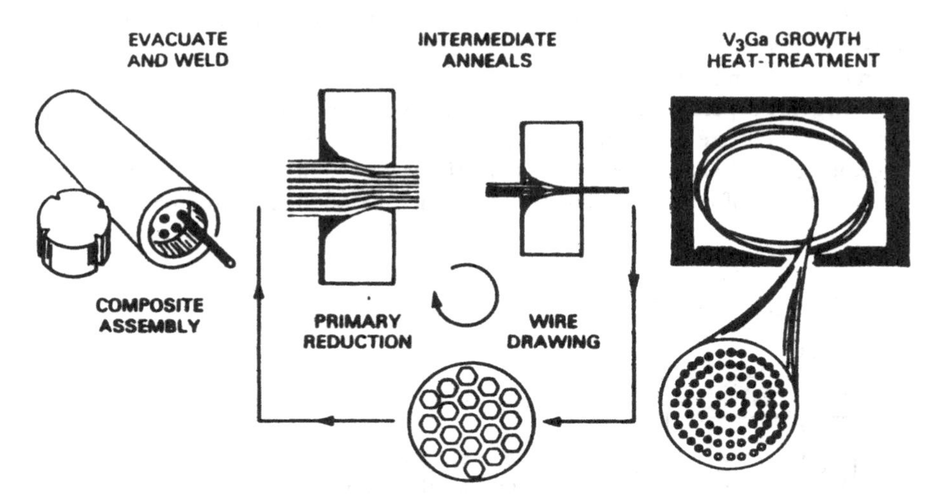

Figure 9.2. Process for manufacturing A15 structure superconducting multifilamentary wire.

several microns of the A15 compound. These superconducting filaments are then surrounded by bronze and copper sheaths, forming the stabilizing matrix for the superconductor. Magnets made from these wires can produce magnetic fields of 20 T at 4.2 K.

B1 COMPOUNDS

Another class of superconducting compounds have the B1 crystal structure shown in Fig. 9.3. This structure is the NaCl face-centered cubic structure with alternating A and B elements in all directions. Several compounds with this very simple crystal structure have high T_c. For example, $NbN_{0.92}$ has a transition temperature of 16 K. The compound cannot be prepared with 1/1 stoichiometry, so the structure has many vacancies. If these vacancies are filled by carbon to form $NbC_{0.1}N_{0.9}$, then the T_c becomes 18 K. As opposed to the A15 materials, $NbC_{0.1}N_{0.9}$ has a high density of states at the Fermi level, but the band is very flat so that disorder does not drastically change the density of states as the Fermi level changes. Thus, the B1 materials are far more resistant to radiation damage and disorder than the A15 materials. Nonetheless, these materials are also quite brittle and are difficult to fabricate into wires, but they have been successfully exploited as thin films for superconducting electronics.

ORGANIC SUPERCONDUCTORS

A model of a superconducting organic polymer was proposed by Little in 1964. This model was described in Chapter 6; here we just note that it initiated a search for organic superconductors.

Organic superconductors are an unusual type of conducting materials. Most organic crystals are insulators. For instance, the conductivity of the well-known organic material anthracene is less than $10^{-22}\ \Omega^{-1}\ \mathrm{cm}^{-1}$. There exists, however, a family of organic metals (the first one, the perylene–bromine complex,

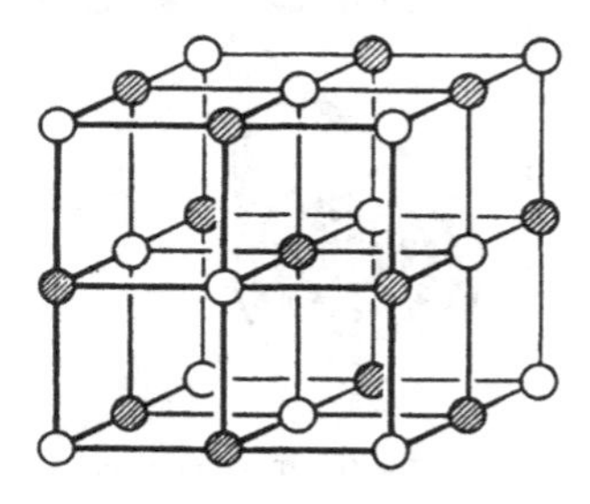

Figure 9.3. The B1 or sodium chloride structure. The solid circles are the A atoms and the filled circles the B atoms.

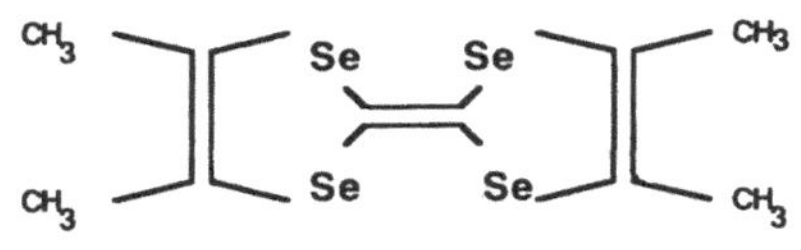

Figure 9.4. The TMTSF molecule.

with a conductivity σ of $\sim 1\ \Omega^{-1}\ \mathrm{cm}^{-1}$, was discovered only in 1954). At present, there are organic metals with conductivities as high as $\sim 10^4\ \Omega^{-1}\ \mathrm{cm}^{-1}$. This conductivity is certainly not very high, but this is not relevant from the point of view of the search for superconductivity, because we know that a high value of the former is not at all favorable for the latter. Such excellent conductors as, say, gold and copper are not superconducting at all. On the other hand, certain metallic ceramics (copper oxides) are not just superconducting, but retain this property up to record high temperatures.

The first organic superconductor was discovered by Jerome *et al.* in 1979, with $T_c \approx 1$ K. However, the ten years following this discovery saw a remarkable increase in T_c. In 1990, an organic superconductor with $T_c \approx 12$ K (Williams *et al.*) was synthesized! The derivative $\partial\ T_c/\partial t$ (where t is the time) is quite impressive.

At present, there are two known classes of organic superconductors. One of them (the so-called Bechgaard salts) is described by the chemical formula $(TMTSF)_2X$. Tetramethyltetraselenafulvalene (TMTSF) is shown in Fig. 9.4; X is a monovalent inorganic anion. Typical anions are PF_6^-, AsF_6^-, NbF_6^-, and $C10_4^-$.

The other class is formed by materials with the composition $(BEDT\text{-}TTF)_2X$, abbreviated $(ET)_2X$. The bis(ethylenedithio)tetrathiafulvalene (BEDT–TTF, or ET) molecule is shown in Fig. 9.5; X is again a monovalent anion. This class is characterized by higher critical temperatures. For example, the crystal $(ET)_2I_3$ remains superconducting to $T_c = 8.1$ K.

Speaking of the physical properties of organic superconductors, one should first of all note their strong pressure dependence. For example, the $(TMTSF)_2C10_4$ material has $T_c = 1.2$ K at ambient pressure, but if we use a Pf_6^- or an AsF_6^- anion instead of $C10_4^-$, the superconductivity (also with $T_c = 1.2$ K) appears only under external pressure exceeding 8 kbar. In other words, when the former two anions are present, the phase diagram shown in Fig. 9.6 obtains.

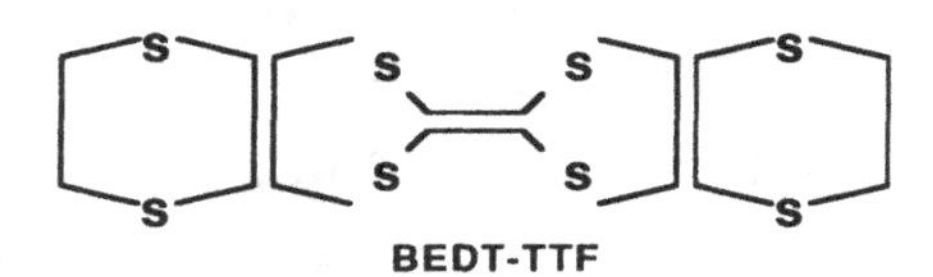

Figure 9.5. The ET molecule.

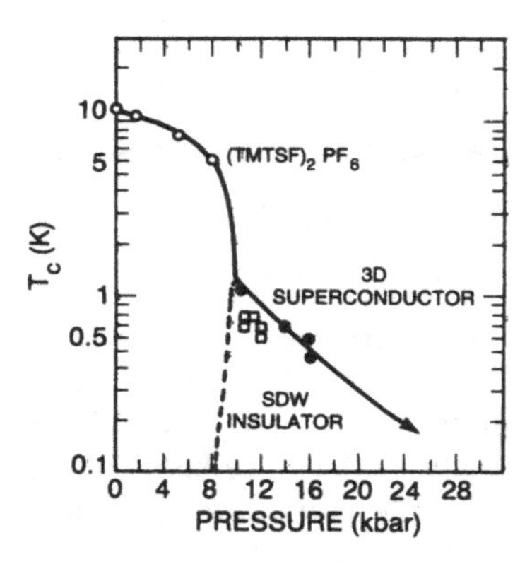

Figure 9.6. *P–T* phase diagram of $(TMTSF)_2PF_6$ or $(TMTSF)_2AsF_6$.

Another characteristic feature of organic metals and superconductors is their very high structural anisotropy. This can be immediately seen from the structure of $(ET)_2BrO_4$ shown in Fig. 9.7. There is a large difference between the resistivity in the plane of the crystal, $\rho_{\parallel}$, which corresponds to metallic behavior), and that in the perpendicular direction, $\rho_{\perp}$, which is of a nonmetallic character. As a result, this material behaves like a quasi-two-dimensional metal.

It is also curious that organic superconductors with the same chemical formula can exist in a variety of crystal phases. For example, there are at least five known phases of the material $(ET)_2I_3$ that differ considerably in their critical temperatures. Even for a single phase [so-called β-$(ET)_2I_3$], the value of T_c varies depending on the method of preparation. Chemical synthesis results in $T_c \approx 1.5$ K, but if the sample is obtained by special vacuum sublimation of iodine, then T_c moves sharply higher, to 8 K. This latter method also results in the samples being spatially inhomogeneous, as manifested by T_c becoming dependent on the procedure used to measure it. The diamagnetic transition takes place at temperatures considerably lower than the resistivity transition.

We have already mentioned that tunneling spectroscopy is a unique method which makes it possible to elucidate the origin of the superconducting state in a given material. Recently, tunneling experiments on the organic superconductor

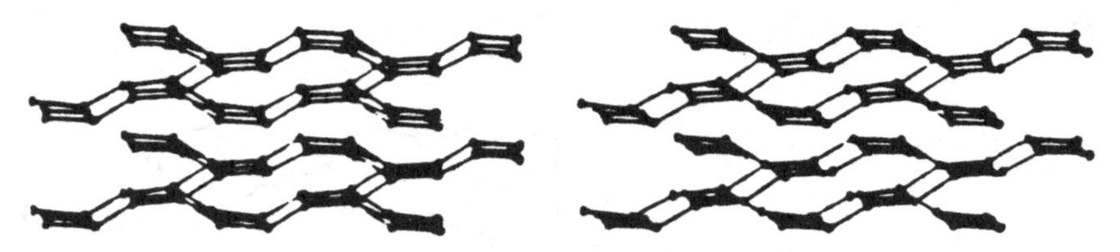

Figure 9.7. Anisotropic structure of $(ET)_2BrO_4$.

β-$(ET)_2AuI_2$ with T_c = 3.8 K were carried out (Hawley *et al.*, 1986). The measured value of the ratio $\Delta(0)/T_c$ corresponds to strong coupling. This may be due to the contribution of the low-lying vibrational modes. However, on the whole, the origin of superconductivity in organic materials is still an open question and is being intensively studied. Some NMR data suggest the possibility of triplet pairing; consequently, the role of magnetic interactions is important. The possibility of an important contribution from electronic excitations also cannot be ruled out.

Organic superconductors are in many respects similar to the cuprates (reduced dimensionality, low E_F, see Chapter 13 and Appendix E). However, the former are distinguished by their strong in-plane anisotropy. If one succeeded in reducing the effect of nesting states, this could produce further rise in T_c (see Wolf and Kresin in Kresin and Little, 1990).

The great variety of possible configurations and chemical compositions, the underutilized potential of chemical synthesis, together with the impressive growth of T_c in recent years combine to make the field of organic superconductivity highly promising.

HEAVY FERMIONS

The family of heavy-fermion superconductors is a contemporary of the organic materials: they also were discovered in 1979 (Steglich *et al.*, 1979).

Heavy fermions are not at all distinguished by high values of T_c; on the contrary, their critical temperatures are rather low. For example, $CeCu_2Si$ has T_c = 0.5 K; in UBe_{13}, T_c = 0.85 K; and in UPt_3, T_c = 0.5 K. The principal feature of these materials is reflected in their name: it is the gigantic value of the effective mass, several hundred times greater than that of a free electron.

Among the normal properties of heavy fermions, the large effective mass manifests itself first of all in the large value of the Sommerfeld constant, $\gamma = C/T$. The magnetic susceptibility is also very large.

The superconducting state also displays some anomalous properties. The first one is again related to the behavior of the heat capacity. In conventional superconductors, the electronic heat capacity decreases exponentially with temperature (see Chapter 1). In heavy fermions, on the other hand, a power law decrease is observed instead. For instance, in UBe_{13} such behavior was observed down to $T/T_c \approx 1/13$. An exponential decrease ($\sim e^{-\Delta/T}$) reveals that there is an energy gap near the Fermi surface. The absence of such a decrease indicates that the gap has nodes along some lines on the Fermi surface, or vanishes at some points. This kind of behavior is incompatible with a spherically symmetric (s-pairing) order parameter $\Delta(\mathbf{r})$ (which plays the role of the wave function of a Cooper pair). It appears that the order parameter has a complicated coordinate dependence and is characterized by an intrinsic anisotropy. This indicates that there is pairing with a finite (nonzero) angular momentum. In this case, the total spin of a pair will be different from zero, so that the pairs are in a triplet state.

Similar nonexponential behavior has been observed in sound attenuation. Unfortunately, it has so far been impossible to carry out tunneling spectroscopy experiments on heavy fermions: it is extremely difficult to prepare the thin films required for such measurements.

The experimental data indicate that heavy fermions display an unconventional type of superconductivity. This point of view is supported by an observation of *two* transition points in UBe_{13}. Specifically, heat capacity measurements (Fischer *et al.*, 1989) revealed two jumps at $T_{c1} \approx 0.49$ K and $T_{c2} \approx 043$ K. We are thus dealing with two superconducting phases.

A similar phenomenon is observed in superfluid ^3He (see Chapter 10). In ^3He, pairs are formed with finite momenta; this leads to different symmetries of the order parameter.

One would like to know, what is the mechanism of superconductivity in heavy fermions? There is considerable evidence that it is caused not by electron–phonon interaction as in conventional superconductors, but by exchange of magnetic fluctuations. On the other hand, electron–phonon interaction in these systems may be highly unusual as well. In usual metals, $E_F/\Omega_D \gg 1$, while in heavy fermions the reverse is true. This renders inapplicable the adiabatic Born–Oppenheimer approximation, which forms the basis of the theory of electron–lattice interactions. It is also possible that we are seeing effects due to overlapping bands and multigap structure.

Physics of heavy-fermion materials is a young and exciting field where at present there are many more questions than answers.

SUPERCONDUCTIVITY IN SEMICONDUCTORS

The possibility of observing superconductivity in semiconductors has long attracted a lot of attention, not just from the point of view of increasing the number of superconducting materials, but because semiconductors possess a number of interesting properties.

The carrier concentration in semiconductors is lower than in metals; from the point of view of the usual BCS theory, this is not a favorable feature. Indeed, according to the BCS model, the coupling constant λ is equal to $V\nu_F$ [see also Eq. (6.16)]. Here ν_F is the density of states, which depends on the carrier concentration n: $\nu_F \sim m^*p_F \sim m^*n^{1/3}$. Thus, one can expect that the coupling constant in degenerate semiconductors will be small. These "doped" materials have a Fermi surface, but the carrier concentration is low. Nonetheless, it is also known that semiconductors are usually characterized by large dielectric constants

ϵ. This considerably weakens the Coulomb repulsion which would otherwise oppose the interelectron attraction. Consequently, the possibility of superconductivity in semiconductors cannot be excluded.

A detailed theoretical analysis was carried out by Cohen (in 1963) (see Cohen, in Parks, 1968), who pointed out the particular role of the so-called intervalley transitions. The Fermi surface of many semiconductors (e.g., $SrTiO_3$) is made up of a set of disjoint regions ("valleys"). Intervalley carrier transitions, induced by the electron–phonon interaction, involve large momentum transfers and are quite effective.

The first superconducting semiconductor to be discovered was GeTe ($T_c \approx$ 0.1 K; Hein *et al.*, 1964) (Parks, 1968). This was followed by the discovery of superconductivity in $SrTiO_3$ ($T_c \approx 0.3$ K; J. Schooley *et al.*, 1965) (Parks, 1968). This material is an oxide with a perovskite structure.

At the beginning of this section we said that, according to the BCS model, low carrier concentration is a negative factor. The real situation is more complicated, however, and increasing the number of carriers does not always increase the critical temperature. It is curious to note the peculiar concentration dependence of T_c in semiconductors. Figure 9.8 shows this dependence for $SrTiO_3$. It is characterized by a sharp maximum; further increase in concentration leads to a drop in T_c.

*The nonmonotonic dependence of T_c on n can be explained with the help of an expression for the coupling constant (Kresin, 1971). Consider Eq. (6.17). If the carrier concentration is low, the Fermi momentum $p_F \sim n^{1/3}$ is relatively small, so that $2p_F < q_c$. Also, $k_1 = 2p_F$, and, in addition, $\Omega(q) = uq$. Then $\lambda \sim \nu \sim Vmp_F$, as in the BCS formula (Eq. 2.6). In this case, $\lambda \sim n^{1/3}$ and increases (as does T_c) with n. Further increase in n leads to a crossover, so that $2p_F > q_c$.

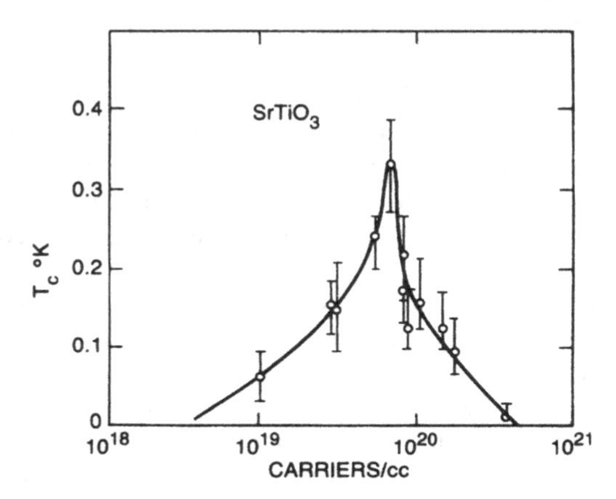

Figure 9.8. Dependence of T_c on carrier concentration.

Then $k_1 = q_c$, and $\lambda \sim p_F^{-1}$. In this case, T_c decreases with increasing n. When $2p_F = q_c$, there is a jump in the derivative (see Fig. 9.8). Note that if the dispersion law changes with increasing q, this may also lead to the appearance of $T_c^{\max}$.*

Note also that the recently discovered high-temperature superconductors also have low carrier concentrations. Thus, small n does not by itself mean that T_c will be low: there may be deviations from the simplified model with $\lambda \sim p_F$.

Speaking of superconducting semiconductors, one should also point out that external pressure can have a strong influence. A study carried out by M. L. Cohen, G. Martinez, and their colleagues in Berkeley and Grenoble has led to a record critical temperature for nontransition elements. At a pressure of 15.2 GPa, they observed $T_c = 8.2$ K for Si.

OXIDES AND HYDRIDES

Superconducting oxides are a special family of superconductors. They turned out to be the ones with high critical temperatures. High-temperature superconducting oxides are a separate topic; see Chapter 13.

The first superconducting oxides were NbO (Miller *et al.*, 1965) and the aforementioned $SrTiO_3$ (Schooley *et al.*, 1965) (see also Cohen, in Parks, 1968). The critical temperatures of these compounds are rather low (1.2 and 0.8 K, respectively). An exceptionally interesting material is $BaPb_{1-x}Bi_xO_3$ (Sleight *et al.*, 1975) with $T_c \approx 12$ K (for $x \approx 0.25$). This T_c has appeared mysteriously high ever since the discovery of this material. Other members of this family are $BaPb_{0.75}Sb_{0.25}O_3$ ($T_c \approx 0.3$ K) and $Ba_{0.6}K_{0.4}BiO_3$ (discovered in early 1988; $T_c \approx 29$ K!). If this latter material had been discovered two years earlier, it would have been a record holder, and its discovery a sensation.

There is thus a family of oxides (we are not talking about the copper oxides) with a very wide range of critical temperatures. This family, and $Ba_{0.6}K_{0.4}BiO_3$ in particular, is currently being intensely studied, and so far the number of questions exceeds the number of answers. It is quite possible that superconductivity in these materials is due to electron–phonon interaction with a peculiar distribution of the phonon spectrum. However, one cannot exclude other possibilities.

Let us briefly discuss materials containing another "familiar" element: hydrogen. Here first of all we should mention the search for metallic hydrogen. Under normal circumstances, solid hydrogen is a dielectric. However, under extremely high pressures, when hydrogen atoms will come close together and electrons will be able to jump from one atom to another and to move freely throughout the crystal, solid hydrogen should become a metal, and possibly a superconductor. It would be the lightest metal in nature. Possibly, it would also have a high T_c.

An interesting system is the palladium–hydrogen or palladium–deuterium compound. Palladium by itself is not a superconductor, but its hydride Pd—H is, with T_c strongly dependent on the degree of hydrogen saturation. The highest T_c (9 K) is observed when H/Pd $\approx$ 1 (Stritzker and Buckel, 1972). These authors also have replaced hydrogen by deuterium and observed a negative isotope effect: the critical temperature for Pd—D turned out to be 11 K (we mentioned this phenomenon in our discussion of the isotope effect in Chapter 6). This interesting system deserves further study.

10

THE SUPERCONDUCTING STATE IN NATURE

We have already remarked that superconductivity is not just the absence of electrical resistance in some solid conductors. In fact, we are dealing with a peculiar correlated state of matter. In this chapter we will talk about the properties of a number of objects which, at first glance, have nothing in common. However, the real subjects of discussion will be the interconnection of different natural phenomena and the various manifestations of the peculiar superconducting state of matter. Low-temperature physics in general, and the physics of superconductivity in particular, has had an enormous influence on other branches of physics. As we shall see, many examples of this influence were quite unexpected, which makes the picture all the more interesting.

SUPERFLUIDITY OF LIQUID HELIUM

Quantum Fluid

Liquid helium is often called a quantum fluid. This name reflects the uniqueness of its properties. As is well known, as a substance is cooled, it goes from the gas phase into the liquid phase, and then solidifies. Helium is the only substance in nature which at ordinary pressure does not solidify even at the absolute zero of temperature.

From the point of view of classical physics, such behavior is totally inexplicable. Indeed, according to classical mechanics, at absolute zero, that is, in the state of minimum energy, the particles making up the crystal should be completely at rest. This picture corresponds to the solid ordered state of matter. It is only at finite but low temperatures that the ions should execute small vibrations about their equilibrium positions.

The extraordinary behavior of liquid helium is explained by the laws of quantum mechanics. According to quantum mechanics, substances are not obliged to become solid when cooled. In fact, at absolute zero, the particles are not at rest, but are performing so-called "zero-point" vibrations.

The physical origin of these zero-point vibrations lies in the Heisenberg uncertainty relations, which are of fundamental importance in the quantum theory and have the form $\Delta p_x \Delta x \approx \hbar$, $\Delta p_y \Delta y \approx \hbar$, $\Delta p_z \Delta z \approx \hbar$, where Δx, Δy, Δz and Δp_x, Δp_y, Δp_z are the uncertainties in the values of particle coordinates and momenta, respectively. It is impossible to simultaneously specify the position and the momentum of a microscopic particle: there cannot be a state in which $\Delta p_x = 0$ and $\Delta x = 0$. Consider a particle performing small oscillations of frequency ω about the equilibrium position. As is known from ordinary classical mechanics, its energy is given by

$$E = \frac{p^2}{2m} + \frac{m\,\omega^2\,x^2}{2} \tag{10.1}$$

This particle is called a harmonic oscillator; the ionic system of a crystal represents an assembly of harmonic oscillators.

From the above equation, one can see that a classical oscillating particle has zero energy only if $p = 0$ and $x = 0$. In this case, both the position and the momentum have definite values. This is impossible in quantum mechanics, and, as a result, a harmonic oscillator must have a finite minimum energy. This energy turns out to equal $\hbar\omega/2$. In other words, an oscillating microscopic particle "cannot stop." The minimum-energy state corresponds to zero-point vibrations with energy $\hbar\omega/2$.

Thus, even at absolute zero the ions in a crystal are not at rest, but are performing zero-point vibrations. If the energy of this zero-point motion is sufficiently high, that is, if the vibrations are intense enough, there will be no transition into the solid phase at all. This is precisely what is observed in liquid helium. All other substances manage to solidify before quantum effects prevent it.

One can wonder why helium turns out to be such an exceptional substance in which quantum effects play such a decisive role. The explanation is connected with two factors: the inertness of helium atoms and their lightness. The inertness leads to the binding forces, which are needed for solidification, being very weak. Besides, since helium is one of the lightest elements, the energy of its zero-point vibrations is quite large (as usual, the frequency is inversely proportional to the square root of the particle mass). Thanks to these facts, physicists have in their possession a quantum fluid, that is, a macroscopic substance whose behavior is governed by the laws of quantum mechanics.

Helium can be turned into a solid by subjecting it to high pressures. Then helium atoms come close together and the binding forces begin to dominate. At temperatures close to absolute zero, it takes a pressure of about 25 atm to obtain solid helium; at $T = 60$ K, it takes about 10,000 atm.

The very existence of a substance that remains liquid down to absolute zero at ordinary pressures represents a striking demonstration of the laws of quantum physics.

Two isotopes of helium exist: ${}_2^4$He, whose nucleus consists of two protons and two neutrons, and a lighter isotope, ${}_2^3$He, which contains only one neutron. Helium found in nature consists primarily of the heavier isotope. Advances in nuclear physics have made it possible to obtain sufficient quantities of ${}_2^3$He as well, so in fact there exist two quantum fluids which have quite different properties.

${}_2^4$He atoms, which contain an even number of particles (nucleons and electrons), are bosons. On the other hand, the light ${}_2^3$He atoms are fermions. This fact is responsible for the differences in the properties of the two quantum fluids. Nevertheless, they have one property in common: both are superfluids. The phenomenon of superfluidity is analogous to superconductivity. In superconductors, the charged electron fluid moves through the crystal lattice without exchanging energy with the latter and, as result, without experiencing any resistance. Similarly, a superfluid moves through a capillary without resistance. That is to say, the viscosity (friction) is zero.

We will give a brief overview of the properties of both quantum fluids. We would like to remark that the theory of superconductivity was especially important in analyzing the properties of ${}_2^3$He. This is naturally related to the fact that this isotope and the electrons in a metal obey the same type of statistics.

Superfluidity of He II

Helium becomes liquid at the remarkably low temperature of 4.22 K (at atmospheric pressure). This is due, of course, to the weak interaction between inert helium atoms. Afterwards, helium remains liquid all the way down to absolute zero (again, at atmospheric pressure).

This does not exhaust the remarkable properties of this substance. At the temperature of 2.19 K, helium undergoes a second-order phase transition. In 1932, Keesom and Clausius discovered an anomaly in the heat capacity near this temperature (Fig. 10.1). The graph of the heat capacity in this region looks like the Greek letter, λ, and so the phenomenon was termed the λ-effect, and the temperature at which it took place was called the λ-point. With increasing external pressure, the λ-point moves toward lower temperatures.

Thus, liquid helium can be in one of two phases separated by the λ-point. These phases are called He I (above 2.19 K) and He II (below the λ-point). He I is an ordinary liquid, but the situation is completely different in the low-temperature phase. He II possesses very anomalous physical properties; it is in a

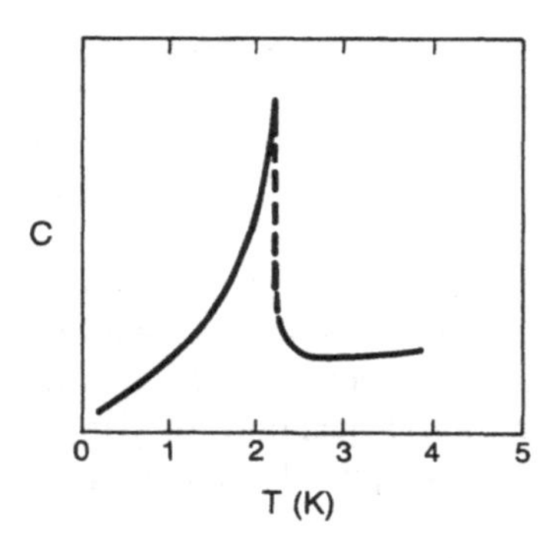

Figure 10.1. λ Transition.

peculiar, superfluid state. In the superfluid state, discovered by Kapitza in 1938, liquid helium is capable of flowing through thin capillaries without any friction.

Superfluidity is 27 years "younger" than superconductivity. However, whereas almost half a century passed between the discovery (1911) and the theoretical explanation (1958) of superconductivity, the history of superfluidity is different. The nature of this phenomenon became clear just three years after the discovery, when in 1941 Landau formulated the theory of superfluidity.

Landau's theory is based on the conjecture that any weakly excited state of a many-body quantum system can be visualized as composed of a set of elementary excitations, or quasiparticles. The characteristic features of the quasiparticle method (its basic principles were established by Landau in working out the theory of quantum fluids) are discussed in Appendix B.

Elementary excitations are characterized by definite values of energy and momentum. It should be stressed once more that by no means should quasiparticles be identified with the real helium atoms. They are associated with the fluid as a whole; the energy of an elementary excitation corresponds to the energy level of the entire quantum system, which consists of an enormous number of particles.

There is a definite dispersion relation between the energy of a quasiparticle and its momentum. Figure 10.2 shows the spectrum of elementary excitations which Landau postulated for He II.

In the beginning part of the spectrum, there is a linear relation between energy and momentum. It corresponds to low momenta, or to long-wavelength excitations of liquid helium. The linear relation means that we are dealing with an ordinary acoustic branch of the excitation spectrum, associated with sound quanta (phonons).

The acoustic character of this initial part of the spectrum means that at low temperatures the excited states of helium are sound waves. Let us address this point in more detail. At high temperatures, helium is a gas, and the individual

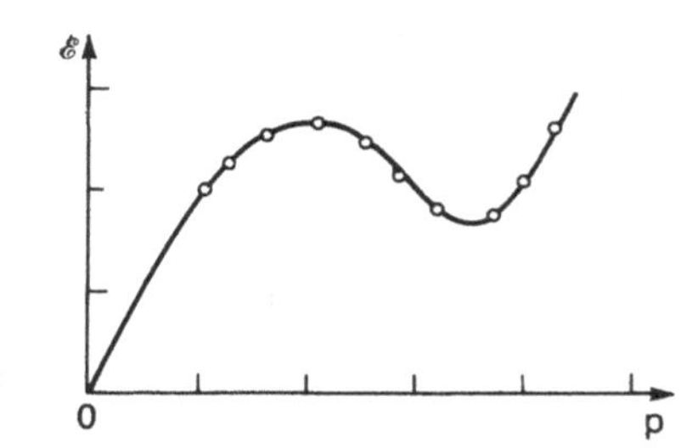

Figure 10.2. Spectrum of He II. Solid line: theoretical curve; ○: neutron spectroscopy data.

gas atoms perform chaotic thermal motion. This, of course, applies to any gas. At lower temperatures, the substance becomes a liquid. This state is also characterized by thermal chaotic motion of individual particles. Sound waves can propagate in a liquid, but they are quickly damped. He I is an example of an ordinary liquid.

Further decrease in temperature is usually accompanied by solidification. In this process, particle motion becomes collectivized. The ions in a crystal oscillate about their equilibrium positions, but these thermal vibrations are then transferred to other particles, and a collective type of motion (sound waves) results.

Thus, at temperatures close to absolute zero, chaotic motion is replaced by collective sound wave motion.

Under ordinary conditions, helium does not become a solid when cooled. However, at near-zero temperatures, motion becomes collective here as well. The "freezing out" of thermal motion leads to sound waves no longer being damped. They become the elementary excitations of the liquid He II quantum system.

These elementary excitations of He II, sound quanta, correspond to the initial linear part of the spectrum in Fig. 10.2.

Based on the main principles of the method of elementary excitations and on the features of the He II energy spectrum, Landau showed that a quantum fluid will indeed be a superfluid. That is, there will indeed exist a type of motion with no friction between the moving fluid and the walls.

Consider liquid helium flowing in a capillary with velocity $\mathbf{v}$ at $T = 0$ K. Friction would lead to the moving fluid losing kinetic energy and slowing down. However, the energy of liquid helium, which is a system governed by the laws of quantum mechanics, cannot change in arbitrary steps. Its excitation corresponds to the appearance of a quasiparticle with energy $E(p)$. [$E(p)$ is the quasiparticle energy in the frame of reference tied to the moving helium.] In the stationary frame (i.e., with respect to the capillary walls), the energy of the system will change by an amount $E + \mathbf{p}\cdot\mathbf{v}$. If we want this change to be energetically favorable, it must be negative; that is, we must have $E + \mathbf{p}\cdot\mathbf{v} < 0$. The left-hand side is minimal if **p** and **v** are antiparallel. However, even in this case, the following must hold:

$$E - p \cdot v < 0 \tag{10.2}$$

This does not hold for velocities

$$v < E(p)/p \tag{10.3}$$

The physical meaning of this equation is as follows. As mentioned above, the elementary excitations in He II are sound waves. Suppose He II is flowing through a capillary at a speed v relative to the walls. Let us go over into the reference frame in which helium is at rest. In this frame, the walls are moving with respect to the helium with velocity $-v$. What is the condition for the wall motion to create an excitation in helium? (If this happens, the translational energy will be dissipated; that is, viscosity will appear.) If the wall velocity exceeds the speed of sound, $v > u$, a sound wave will be generated. This is the well-known phenomenon of Cherenkov radiation: an object moving through a medium faster than the velocity of wave propagation will emit waves.

If $v < u$, no waves will be emitted. The condition given by Eq. (10.3) corresponds precisely to this case (for small p, the ratio E/p is equal to the speed of sound; see Fig. 10.2). Therefore, if the helium flow velocity satisfies Eq. (10.3), no sound waves will be excited. In this case, helium does not become excited when moving through a capillary, which means that the flow is resistanceless, that is, superfluid.

We should point out that the condition given by Eq. (10.3) will also be satisfied when the spectrum has a gap at $p = 0$: $E(p = 0) = \Delta$. Then the flow will be dissipationless for $v < (\Delta/p)$. This corresponds to the usual superconductivity, and also to the superfluidity of ${}^3_2He$ (see below).

In the two-fluid model (a similar model has also been successfully used in the physics of superconductivity), He II is viewed as consisting of two components: normal and superfluid. This concept permits a clear visualization of the basic properties of superfluids. The fluid density ρ is written as $\rho = \rho_n + \rho_s$, where ρ_n is the density of the normal component, whose properties are identical to those of He I, and ρ_s is the density of the superfluid component. When $T \rightarrow 0$, $\rho_n \rightarrow 0$, and the entire liquid becomes superfluid. On the other hand, when the λ-point is crossed, the superfluid component vanishes. The two components are assumed to move freely through each other without any friction. The superfluid component has zero entropy and does not transport any heat at all. It is this component that sustains frictionless flow and thus leads to superfluidity.

It should be stressed, though, that just as in the treatment of superconductivity, the two-fluid model is only an intuitive picture designed to visualize the complicated behavior of a quantum fluid. No classical model can adequately reflect the quantum-mechanical picture. In reality, the fluid is not partitioned into two distinct components. A given helium atom can be neither normal nor superfluid. It is more correct to speak not of two components, but of two possible types of motion of a quantum fluid: normal and superfluid.

The two-fluid model allows many exotic phenomena observed in He II to be explained. For instance, an object can move through the superfluid component without any resistance. In this regard, the behavior of the superfluid component resembles that of an ideal liquid, studied in hydrodynamics. The normal component does not have this property. This difference forms the basis for the following curious effect. If one sets up a temperature gradient in a tube filled with liquid He II, there will arise a counterflow of the normal and superfluid components. An object immersed into the liquid will feel a force exerted by the normal constituent and will begin to move, even though there is no real mass flow of the liquid. This was observed by Kapitza in one of his experiments.

If a vessel filled with liquid He II is rotated, the observed moment of inertia will be less than that observed when a usual liquid is rotated. This is due to the fact that the superfluid component is not dragged along by the rotating vessel. In this way, one can measure the relative concentrations of the normal and superfluid components.

The Landau theory is based on the He II excitation spectrum that he postulated. It is possible to measure this spectrum directly by studying neutron scattering in liquid helium. Neutrons impinging on helium undergo inelastic scattering because they create He II elementary excitations. Conservation of energy and momentum requires that

$$\frac{p_1^2}{2m} - \frac{p_2^2}{2m} = E(p)_i \qquad \mathbf{p}_1 - \mathbf{p}_2 = \mathbf{p} \tag{10.4}$$

Here $\mathbf{p}_1$ and $\mathbf{p}_2$ are the initial and final neutron momenta, respectively; $\mathbf{p}$ and $E(p)$ are the momentum and energy of a quasiparticle corresponding to helium excitation. These relations allow $\mathbf{p}$ and $E(p)$ to be determined from measurements of the energy and momentum of a neutron scattered at a known angle, thus yielding the He II excitation spectrum. Figure 10.2 illustrates the results of such measurements, which have confirmed the theory.

A theoretical calculation of the spectrum in Fig. 10.2 is extremely complicated. It requires a quantum-mechanical analysis of the real liquid helium, accounting for interatomic interactions. These interactions are not weak, and this makes the calculation of the He II excitation spectrum very difficult. In 1947, Bogoliubov considered a microscopic model of a nonideal Bose gas made up of weakly repelling particles. He showed that for low momenta the excitation spectrum of this system was made up of sound waves. Therefore, the model exhibits superfluidity.

A very important phenomenon in the study of Bose gases is the Bose–Einstein condensation. In 1924, Einstein pointed out a peculiar property of an ideal Bose gas, that is, of a system of noninteracting identical particles of integer spin. At $T = 0$ K, all the particles accumulate in the state of zero momentum. This is the state of the lowest possible energy, and since the particles have integer spin, the Pauli principle does not prevent them from all gathering in this state. Of course, at finite temperatures, thermal motion leads to the occupation of other states as well. However, an important fact is that at temperatures close to the absolute zero and less than a certain characteristic temperature T_0, there is still a finite number of particles in the $p = 0$ state. Only for $T > T_0$ does an infinitely small momentum interval dp near $p = 0$ begin to contain an infinitely small number of particles, as is usual. This means that at temperatures less than T_0 (which is called the condensation temperature), bosons will exhibit ordered behavior. It is important that this ordering takes place in momentum space rather than in coordinate space.

Of course, real helium is not an ideal Bose gas, but there is still reason to believe that the superfluid transition is related to ordering, that is, to the appearance of a macroscopic number of particles having $p = 0$. This is indicated, first of all, by the fact that the He II superfluid transition temperature ($T = 2.19$ K) is very close to the ideal Bose condensation temperature, $T_0 = 2.17$ K (if the boson mass is taken to be that of a helium atom). However, the Bose condensate, that is, the agglomerate of particles with zero momentum, should not be confused with the superfluid component of liquid helium. For one thing, this follows from the fact that at $T = 0$ K the mass of the superfluid component becomes equal to the mass of the whole liquid, while only a small fraction of particles will be at rest ($p = 0$), because of the strong interatomic interaction. Neutron scattering experiments (the small de Broglie wavelength of the neutrons allows them to interact with individual helium atoms) have shown that this fraction is only about 7%.

Note that superconductivity is caused by electrons forming Cooper pairs, which are bosons of a sort; this illustrates once more the unity of the phenomena of superconductivity and superfluidity.

It is beyond the purpose of this book to give a detailed account of the properties of liquid helium. We shall give only a brief description of some additional phenomena that illustrate the analogy between superconductivity and superfluidity. One should not, of course, expect a full analogy, because there are some fundamental differences as well. First of all, the electronic system in a superconductor consists of charged particles. In addition, helium atoms are much heavier than electrons, which strongly affects the phase transition. Nevertheless, there are a number of significant analogies. The most interesting is observed in studying the so-called critical phenomena.

As is well known, the superconducting state can be destroyed by an external magnetic field. Obviously, a magnetic field will not have such an effect on superfluid helium, because the latter is not charged. However, if we recall from electrodynamics that a magnetic field is equivalent to a rotating coordinate system, it becomes obvious how one should develop the analogy. Indeed, consider liquid helium in rotation. It turns out that in this case, just as in superconductors, there will arise a vortex structure. In fact, vortices were first introduced and investigated in the study of liquid helium (by Onsager in 1949 and Feynman in 1955) (see Gorter, 1964).

While the rotation is slow, the superfluid component of He II behaves like an ideal fluid and, as mentioned above, does not take part in the rotation. It is at rest in the vessel, while the latter slowly spins around its axis. However, as the speed is increased, there appear vortex lines parallel to the axis of rotation. Near each line, the superfluid component rotates with velocity v determined from the quantization condition

$$mv_s r = nh \tag{10.5}$$

where m is the mass of the helium atom and r is the distance from the vortex axis. This condition is analogous to the Bohr quantization rule. However, as opposed to an atomic orbit ($r \approx 10^{-8}$ cm), the quantity r appearing in Eq. (10.5) can assume gigantic (by quantum standards) values, up to the dimensions of the vessel ($r \approx 1$ cm). The scale of quantization thus may increase by a factor of 10^8 over the atomic scale.

One circulation quantum $v_s r$ is equal to h/m. It was experimentally determined by Vinen (1961), who measured the oscillation frequency of a thin wire surrounded by rotating helium.

According to Eq. (10.5), the velocity distribution in rotating helium is given by $v_s = hn/rm$. This dependence on r is radically different from that observed in classical liquids. If a usual liquid rotates as a whole, then the velocity is determined by the well-known law $v = \omega r$; that is, it increases with increasing distance from the axis of rotation. In a quantum fluid, the velocity field is completely different. As seen from Eq. (10.5), the velocity increases as one approaches the vortex axis.

A vortex line possesses a certain amount of energy. It can be formed, then, only for a definite finite angular velocity Ω_c of helium rotation. The critical velocity is given by the following simple expression: $\Omega_c = (\hbar/m_{He}R)^1 \ln(R/a)$, where R is the radius of the cylindrical vessel containing the liquid helium, and a is on the order of the interatomic distance. Substituting the numerical values of h and m_{He}, we obtain $\Omega_c(\mathrm{s}^{-1}) \approx 1.4 \times 10^{-4} R^{-1} \ln(R/4 \times 10^{-8})$, for R in cm.

From this formula, we see that even for small cylinders the quantity Ω_c is rather small. For example, for R ≈ 1 cm, $\Omega_c \approx 10^{-3}\ \mathrm{s}^{-1}$. Usually, $\Omega \gg \Omega_c$; then the helium contains many vortices.

The formation of vortex lines plays a very important role in the question of the critical velocity of superfluid helium. As described above, in rotating helium these are straight lines parallel to the cylinder axis, and every line has a definite value of angular momentum associated with it.

Vortex formation turns out to be important for studying the flow of helium through a long capillary (Geilikman, 1959). Here the vortex lines are closed curves lying in the plane perpendicular to the capillary axis. Their shape is close to that of the capillary cross section. In this case, what is nonvanishing is not the angular momentum, but the linear momentum of a vortex line, directed along the flow. It is this momentum that is quantized when helium flows through the thin capillary. The critical flow velocity turns out to depend on the cross-sectional shape of the capillary. For example, for a rectangular cross section with sides a and b and $b \ll a$, $v_c = (h/mb)[\ln(2b/d) + 1/4]$; here d is the diameter of the vortex loop, on the order of atomic dimensions. For a round cross section of radius r, $v_c = (h/mr)\ln(r/d)$.

The presence of vortex lines leads to the appearance of friction. Indeed, their presence affects the normal component of He II as well. The quasiparticles corresponding to the normal component scatter off the vortex lines and transfer momentum to them, and consequently to the entire superfluid part. Physically, this means that there is friction between the normal and superfluid components of He II. They no longer move freely and not interact, as was the case before vortex motion appeared. The friction which arises in this way leads to the superfluid component losing energy, that is, to breakdown of superfluidity.

In thick capillaries, where vortex lines form quite easily, the critical velocity is very low. As a consequence, superfluidity is usually studied in thin capillaries.

Superfluidity of $^3_2$He

Everything we have talked about so far has to do with liquid helium made up of $^4_2$He atoms. These atoms contain an even number of particles, have integer spin, and are therefore described by Bose–Einstein statistics.

The abundance of the light isotope $^3_2$He is very low. For instance, in atmospheric helium it makes up only 10^{-7}% of the total. It has become possible to study $^3_2$He only because of the development of methods of producing it artificially. This isotope is usually produced by neutron bombardment of lithium nuclei:

$$^6_3\text{Li} + ^1_0n \rightarrow ^3_1\text{H} + ^4_2\text{He} \tag{10.6}$$

The development of nuclear physics has made it possible to obtain $^3_2$He in large enough quantities to liquefy it. Under normal conditions, this takes place at $T = 3.19$ K.

The properties of this new quantum fluid are quite similar to those of liquid

$^4_2$He above the λ-point, that is, to those of He I. For instance, at $T = 2.5$ K the viscosity of $^3_2$He is $\eta = 1.6 \times 10^{-5}$ poise, while that of He I is $\eta = 3.2 \times 10^{-5}$ poise.

Liquid $^3_2$He is used for making so-called condensation thermometers. It is the only substance that makes it possible to measure temperatures below 1 K with such thermometers. The condensation thermometer is made up of a reservoir filled with liquid which is in equilibrium with its saturated vapor and a manometer for measuring the pressure of this vapor. As is well known, the pressure of saturated vapor does not depend on the volume and is completely determined by the temperature. Therefore, the manometer reading can be used for calibrating the temperature. Helium is the only substance to remain liquid at ultralow temperatures and is therefore used for thermometry near absolute zero.

For physicists who studied liquid helium, the most interesting question was, of course, does liquid $^3_2$He become superfluid? For a long time, this was thought impossible.

The crux of the matter is that the half-integer spin of $^3_2$He atoms (they contain an odd number of particles) leads to an energy spectrum drastically different from that of $^4_2$He. It is natural to consider the spectrum of $^3_2$He to be analogous to that of the electron liquid in a metal, because the latter is also made up of half-integer-spin fermions. In a Fermi system like this, the quasiparticles describing the excited states fill up all the energy levels up to some E_0, in accordance with the Pauli principle. The energy ΔE measured from the Fermi surface can normally be arbitrarily small. Therefore, the Landau superfluidity condition, which in this case is written as $v < E/p$, will not hold. It follows that $^3_2$He should be a normal liquid and experience friction when flowing through capillaries.

The problem of superfluidity of $^3_2$He was radically reassessed after the theory of superconductivity appeared. Soon afterwards, Landau suggested that liquid $^3_2$He, like its heavier isotope, should be a superfluid but that the phase transition should take place at extremely low temperatures, much lower than those at which previous research had been carried out.

As we have discussed extensively, superconductivity is caused by interelectron attraction. The attraction binds electrons in pairs, and this creates a gap in the energy spectrum. The electron system becomes superfluid; that is, it moves through the crystal without resistance.

There are long-range attractive forces in $^3_2$He as well. They are the familiar van der Waals forces, which are forces of an electrical nature acting between neutral polarizable particles. At large distances, electric dipoles attract each other with a force that goes as r^{-7}, where r is the interparticle distance. At sufficiently low temperatures, the weak van der Waals interaction binds $^3_2$He atoms into pairs similar to the Cooper pairs in superconductors. A gap appears in the energy spectrum, and the liquid becomes superfluid.

After the appearance of the theory of superconductivity, there commenced an intensive search for the new superfluid: ${}^3_2$He. The effect has been observed by Osheroff *et al.* (1972) and Wheatley (1973). It takes place at the temperature of just 0.00265 K under 30 atm pressure. The researchers found two phases of superfluid ^{3}He. At the aforementioned temperature, there is a transition into the superfluid A phase, and at an even lower temperature, there is a transition into yet another phase, called the B phase. The two phases differ in their symmetries. The pairs have nonzero angular momenta.

THE "SUPERCONDUCTING" STATE IN NUCLEI

The atomic nucleus is made up of protons and neutrons. These particles have half-integer spins, just like electrons in metals and ${}^3_2$He atoms. In other words, a nucleus is also a Fermi system. The phenomenon of superconductivity, where a gap appears in the energy spectrum, takes place even if interparticle interactions are weak. In a nucleus, the interaction is very strong, and therefore it is natural to suspect that nuclear matter should be superfluid.

How does the superconducting state of nuclear matter manifest itself? Most certainly, not by the absence of electrical resistance: in a nucleus, this concept has no meaning. There are other clues, such as the way in which a nucleus absorbs radiation. This process turns out to depend strongly on whether the nucleus contains an even or odd number of, say, protons. In the former case, the energy of an incoming photon must be equal to or greater than the binding energy of a nucleon pair (see below). On the other hand, if the number of protons (or neutrons) is odd, there is an unpaired particle left over. This makes it possible for the nucleus to absorb lower (by a factor of 10 or so) frequencies than in the first case.

Another indication is provided by the fact that the experimentally measured nuclear moments of inertia are considerably smaller (by a factor of 2–2½) than the values calculated theoretically with the use of the noninteracting particle model.

Pair correlation leads to the formation of bound proton and neutron pairs in nuclei, similar to the electron pairs in superconductors. They are formed in such a way that the total angular momentum of a pair is zero. In order to excite a system of nucleons, a pair must be broken; this requires a certain amount of energy. This leads to the gap between the ground and first excited states which is observed experimentally in even–even nuclei (where all nucleons are paired).

The usual superfluid and superconducting systems are made up of an enormous number of particles and, for the purposes of most problems, may be assumed to be infinite. The peculiarity of an atomic nucleus lies in there being only a comparatively small number of nucleons. In other words, a nucleus is a "finite Fermi system." As a result, surface effects play an essential role in nuclear physics. They are manifested first of all in the fact that single-particle energy levels in nuclei do not form a continuous spectrum (as in an infinite system) but are discrete. Let E_1 be the spacing between the ground and the first excited energy level (not including pairing correlations). Pairing effects will be important only if the pair correlation energy 2Δ exceeds E_1. In other words, at least a few single-particle levels should fit into the energy interval 2Δ.

We have mentioned the difficulties which arose in calculations of nuclear moments of inertia. These calculations had been done for two limiting models. One of them treated the nucleus as a rotating rigid body; the other was based on hydrodynamics. Both were in poor agreement with experimental data.

The most rigorous theory which gives a correct description of the situation (Migdal, 1959) takes into account the effects of pairing correlations. It directly considers the superfluidity of nuclear matter. The moments of inertia turn out to depend on the pairing energy 2Δ. If the pairing energy were infinite, the moment of inertia would correspond to the rotation of an ideal fluid. In the opposite limit of $\Delta \rightarrow 0$, the picture would be that of a rotating rigid body. In reality, we are dealing with a two-fluid model. The detailed analysis has given results in good agreement with experiments. This example shows how an application of ideas and methods originally developed in low-temperature physics has proven fruitful for solving a major puzzle in nuclear physics.

Measurements of nuclear moments of inertia are analogous to the experiments with rotating liquid helium. The smallness of the moments observed in two so vastly different systems is due to a single profound physical principle, discovered in quantum many-body theory. Attractive forces in nuclei lead to nucleon pairing analogous to the Cooper pairing in metals. These paired nucleons form a Bose system similar to helium. The character of the spectrum of this system is such that it cannot be excited by just any external action. The development of the superfluid model of the atomic nucleus has produced a large number of important results.

SUPERCONDUCTIVITY AND ASTROPHYSICS

Let us now turn to a completely different area of physical science, astrophysics. In the study of astronomical objects, such as stars and planets, we are dealing with enormous scales, and one might think that there is no need to talk about quantum phenomena here. It has been established, however, that in astrophysics one also runs into macroscopic quantization. These effects manifest themselves at high, rather than low, temperatures.

In 1932, just one year after the discovery of the neutron, Landau proposed that there may exist neutron stars. This hypothesis was proven right in 1967, when new stellar objects, the so-called pulsars, were discovered at the Cambridge observatory by Hewish *et al.* (see Hewish, 1968). The discovery of pulsars was the greatest achievement of the new field of radio astronomy. The astronomers discovered pulsed radio signals coming from some source within our galaxy. These mysterious pulses were 1/30,000 s long, with long pauses in between. The duration of the pauses was 1.3372275 s. Apart from the pulsed nature of the signals, what was amazing was the constancy (to the eighth decimal place!) of the pause length.

At first, it looked as if the signals were coming from some sort of extraterrestrial civilization. The scientists called the unknown source "LGM," which stood for Little Green Men; there had been recent reports that this was how the creatures in flying saucers looked. However, after several other sources of pulsed radiation had been discovered, it became clear that the astrophysicists were dealing with an entirely new, but still natural, kind of astronomical object. These sources came to be called pulsars, after their signature pulsating signals.

The discovery of pulsars generated enormous interest. At present, over 400 are known. Let us discuss one in particular. In the astronomical classification, it is known as NP 0532. It was discovered in 1968 in the same area of the sky as the Crab nebula. This pulsar is interesting, first of all, because it emits in the optical region of the spectrum. It "switches on" 33 times a second. It is also interesting that its age is well known. The Crab nebula was created in a supernova explosion that took place in 1054 and was seen and recorded by Chinese astronomers. That fact that supernova explosions are accompanied by pulsar formation is important both for the study of stellar evolution and for the establishment of the nature of pulsars.

The study of pulsars and their properties has led astronomers to the conclusion that they are neutron stars.

As the name indicates, neutron stars are almost entirely made up of neutrons. They have an extraordinarily high density, on the order of 2×10^{14} g/cm^3. This kind of density is typical of nuclear matter.

A few words should be said about the formation of neutron stars. As is well known, the radiation from ordinary stars is due to thermonuclear processes in their interiors. Matter in these stars is in the form of a high-temperature plasma, that is, an electrically neutral gas consisting of electrons and positive ions. With time, the gravitational forces in these systems becomes more and more unbalanced by plasma pressure. The density of matter rapidly increases. The protons in nuclei absorb electrons, forming neutrons, and a neutron star is born.

As we have already mentioned, neutron stars are characterized by nuclear densities; they are similar to gigantic nuclei. Pulsars are very small on the astronomical scale, just ~10 km, but because of their enormous densities their masses are comparable to those of the Sun ($M = 2 \times 10^{33}$ g).

The structure of a neutron star is as follows. The external layer is a solid core, more rigid than any steel imaginable. Since the core is so strongly compressed, its temperature (billions of degrees) is lower than its melting point.

The core surrounds a liquid nucleus which is a neutron liquid with an admixture of 5% protons and 5% electrons. The neutron liquid is in a state analogous to that in an atomic nucleus. In view of the earlier discussion, it is then natural to expect that it will be superfluid (Migdal, 1959).

How is the superconducting state of neutron stars manifested? Here one specific property of the superconducting state is important. As we discussed in Chapter 5, superconductors have very low heat capacity. Owing to this property, neutron stars cool very rapidly; this effect must be taken into account in the study of stellar evolution.

The superfluidity of the neutron liquid forming the pulsar nucleus is analogous to that of ${}^3_2$He. In both cases, the superfluidity is due to the formation of bound pairs, similar to the mechanism of superconductivity in metals.

A pulsar rotates about its axis. As a result, we are dealing with a rotating neutron superfluid. Just as in the case of liquid helium, there arises a vortex structure. The presence of vortices results, again as in liquid helium, in the appearance of friction between the normal and superfluid components, and thus in a certain amount of viscosity. As pointed out in our discussion of liquid helium, the vortex structure leads to angular momentum quantization. This quantization also takes place in a neutron star. In other words, the usual Bohr quantization condition holds.

The spatial scale of quantization in this case is worth a special look. In liquid-helium experiments, the scale is set by the diameter of the rotating cylinder. In the present case, the characteristic distance is the diameter of the neutron star itself. Thus, quantization takes place on a scale on the order of 10 km, an increase by a factor of $\sim 10^5$ over the scale for the case of rotating liquid helium. That is to say, pulsars display macroscopic quantization on a scale 10^{13} (!) times greater than the atomic scale. In other words, the very same Bohr rules that govern the behavior of atomic electrons quantize angular momentum in the enormous cosmic realm as well.

The feature that is common to such disparate objects as atoms, superconductors, superfluid helium, and pulsars is the unusual, extremal state they are in. The discreteness of physical quantities, which is the most striking manifestation of quantum regularities, is displayed either in the microscopic world or at ultralow temperatures or, finally, at high temperatures but under extraordinary pressure. Under these conditions, there is no averaging "washing out" of quantization, as is typical for macroscopic bodies which surround us.

Thus, a system of quantized vortices sets up in a pulsar. The most stable configuration of these vortices is a hexagonal lattice. The density of vortices is expressible in terms of the angular velocity of rotation with the help of the following simple formula: $n = 4M_{\mathrm{He}}\Omega/h$. For example, for the optical pulsar in the Crab nebula, $\Omega = 200\ s^{-1}$. The corresponding vortex density is $n = 2 \times 10^5$ cm^{-2}. The period of the lattice is $a = 2.5 \times 10^{-3}$ cm.

Pulsar pulses are exceedingly regular, but there are sometimes discontinuous jumps in the period, termed "glitches." They are seen, for example, in the Vela pulsar (PSR 0833-45) and the already mentioned young optical pulsar NP 0532 in the Crab nebula. The relative changes in the period are quite small. For the Vela pulsar, the period changes by a factor of only 2×10^{-6}. For the Crab pulsar, the glitches are even smaller, the period changing by a factor of 3×10^{-9}, but the observational precision is so high that they are clearly registered.

The mechanism of glitches is usually explained in terms of the so-called "starquakes." A starquake is a rearrangement of the pulsar core. This is accompanied by a release of mechanical tension. Since a pulsar is rotating, it is somewhat squashed at the poles. A starquake is accompanied by some flattening at the equator and by a rise at the poles. The effective radius decreases, and the rotation speeds up. The glitches observed in the Vela pulsar correspond to radius changes of 0.1 cm; for the Crab pulsar, they are much smaller ($\sim 10^{-3}$ cm).

The superfluid neutron liquid is not dragged along by the core but continues to rotate at its old rate. The rotational velocities of the core and the neutron liquid take a long time to equilibrate. The relaxation time is 2×10^8 s (= 6 years) for the Vela pulsar and 2×10^5 s (= 2.3 days) for the Crab pulsar. The picture is very similar to the behavior of rotating liquid helium.

Changes in the period of rotation of a pulsar may be caused by other mechanisms as well. It is possible to have a vortex state with an excess (metastable) number of vortices. This is observed in liquid helium. When the metastable vortices decay, their angular momentum cannot vanish but is transferred to the entire rotating system. Naturally, this alters the rate of rotation.

There are three distinct times involved in a glitch. The first, denoted by γ, is the time between two successive glitches. For the Vela pulsar, it is equal to 2 years, and for NP 0532, 3 months.

After the jump in velocity, the rotation begins to gradually slow down. This process in not uniform. At first, the velocity drops rather fast, and then more gradually. The time t_0 of the transition to the gradual regime is called the relaxation time. For Vela, it is 3.7×10^7 s (= 1.2 years), and for the Crab pulsar, $t_0 = 6 \times 10^5$ s (= 7 days).

After the glitch, there is also an oscillation in the rotation frequency, also with a characteristic time, t_1 (for PSR 0833-45, t_1 = 7 months, and for NP 0532, t_1 = 4 months).

If the pulsar interior were made up of a normal classical fluid, the relaxation times would be much shorter, on the order of 10^{-7} s. Only if the neutron liquid is superfluid and, in addition, contains a system of vortices can there be relaxation times like those listed.

To conclude this discussion, we would like to quote Freeman Dyson:

> I myself find that the most exciting part of physics at the present moment lies on the astronomical frontier, where we have just had an unparalled piece of luck in discovering the pulsars. Pulsars turn out to be laboratories in which the properties of matter and radiation can be studied under conditions millions of times more extreme than we previously had available to us. . . . the pulsars will, during the next 30 years provide crucial tests of theory in many parts of physics ranging from superfluidity to general relativity.

SUPERCONDUCTIVITY AND THE PHYSICS OF COMPLEX MOLECULES

There exists a very large class of complex molecules with so-called conjugate bonds. Their main building blocks are carbon and hydrogen atoms, and these molecules are often referred to as conjugated hydrocarbons. Figure 10.3 shows some sample structures. These molecules contain large numbers of electrons. The majority (called σ electrons) are located close to the atomic nuclei and are not much different from the ordinary atomic electrons. However, along with the σ system, conjugated hydrocarbons also have so-called π electrons, which are not localized near any particular atom. The π electrons can travel throughout

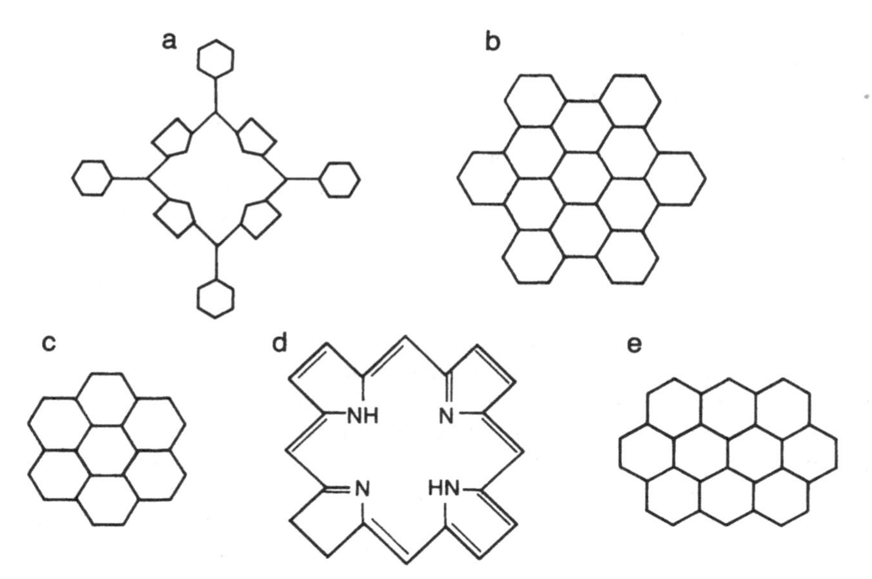

Figure 10.3. Molecules with delocalized electrons: (a) tetraphenylporphin, (b) hexbenzcorenene, (c) corenene, (d) porphin, (e) ovalen.

the entire frame of the molecule. Every carbon atom supplies one such collectivized π electron. Consequently, their number is equal to the number of carbon atoms. For example, the hexbenzcoronene molecule has 32 π electrons, and the chlorophyll molecule has 40.

As we have stated, the π electrons can move throughout the molecular frame. This makes the molecule very similar to a metal. The framework of atoms plays the role of a crystal lattice, and the π electrons that of the conduction electrons.

How is the collective metallic nature of the π electrons manifested? In metals, the mobility of the conduction electrons is displayed when an external electric field induces an electric current. Similarly, the mobile character of the π electrons is most easily seen by studying the behavior of the molecules in an external electric field. Of course, one has to look for a property other than electrical conduction: it makes no sense to talk about the conductivity of a molecule with just a few dozen electrons. If a conjugated hydrocarbon is placed in a longitudinal electric field (the electric field vector is in the plane of the molecule), the molecule will be polarized much more strongly than one that does not contain any collective electrons. Under the influence of the field, the π electrons move to the edge of the molecule and produce a large dipole moment. Clearly, if the electrons were tied to their respective atomic nuclei and could not move throughout the entire molecular core, this large polarizability would not arise.

Thus, a conjugated hydrocarbon with its π electrons is similar to a metal. It turns out, in fact, that it is more than just similar to a metal but is actually a small superconductor (Kresin *et al.*, 1975). The π electrons form bound pairs analogous to the Cooper pairs in ordinary superconductors. The pair correlation mechanism is principally due to polarization of the σ core. Sometimes, a contribution is also made by the interaction of separate groups of π electrons (e.g., in the molecule shown in Fig. 10.3a) similar to the mechanism of superconductivity in the Little model (see Chapter 7).

What are the characteristics of the superconducting state in complex molecules? Whereas the mechanism inducing the superconducting state is similar to the usual superconducting mechanisms, the manifestations of this state are analogous to those of the superconducting state in nuclei. This is not an accident, because in both cases we are dealing with a finite Fermi system, that is, a finite number of particles all governed by Fermi statistics (protons and neutrons in nuclei, π electrons in conjugated hydrocarbons). In both cases, we are dealing with systems with a set of discrete energy levels. The π electrons, their motion being restricted to their molecule, are effectively sitting in a potential well; their energy levels are quantized in the usual quantum-mechanical way.

The superconducting properties of a π-electron system are exhibited primarily (as in nuclei) in the way that the system interacts with radiation. Absorption of light provides information about the excited states of a molecule. It turns out that the energy spacing between the ground state and the first excited state (see Fig. 10.4; absorption of radiation accompanied by a transition from the ground state into the first excited state is called a 0–0′ transition) greatly exceeds the energy of the 0′–0″ transition between the first and second excited states. For instance, in the coronene molecule (Fig. 10.3) the energy of the 0–0′ transition is equal to 22,500 cm^{-1} and that of the 0′–0″ transition is only 5500 cm^{-1}. This disparity is a manifestation of superconductivity in complex molecules. The 0–0′ transition is basically equal to the superconducting energy gap of the π-electron system. In the absence of pairing correlations, from semiclassical considerations one would expect, as in the case of atomic nuclei, a set of roughly equally spaced energy levels.

Interestingly, if one electron is added to the molecule (so that the total number is now odd), the frequency of absorbed light decreases sharply, demonstrating that the 0–0′ spacing has become much smaller. This is accompanied by a dramatic color change of the substance containing the molecules. For example, if one electron is added to the coronene molecule described above, the energy of the 0–0′ transition drops to 8000 cm^{-1}. In a system with an odd number of π electrons, the unpaired electron absorbs light as if there were no superconducting state. If yet another electron is added to the coronene molecule, the 0–0′ transition increases again.

The superconducting state is manifested also in the way molecules interact with an external magnetic field. Usual superconductors display the Meissner effect whereby the magnetic field does not penetrate into the bulk of the material. It makes no sense to talk about the Meissner effect in the molecules we are discussing here since they are flat, but the physical roots of this effect are displayed in conjugated hydrocarbons as well. Experimentally, one observes anomalous diamagnetism of these molecules. In other words, an external magnetic field induces motion of the π electrons which results in an anomalously large reduction, or "repulsion," of the external field. This effect is a manifestation of the superconducting state of the π electrons.

Usually, the total orbital magnetic susceptibility of an electron system is a sum of two terms, $\chi_p + \chi_d$, where χ_p is the paramagnetic and χ_d is the diamagnetic susceptibility. In the well-known quantum-mechanical expression for the current density [see Eq. (5.9)], the first term corresponds to the paramagnetic contribution and the second, which is proportional to the vector potential **A**, to

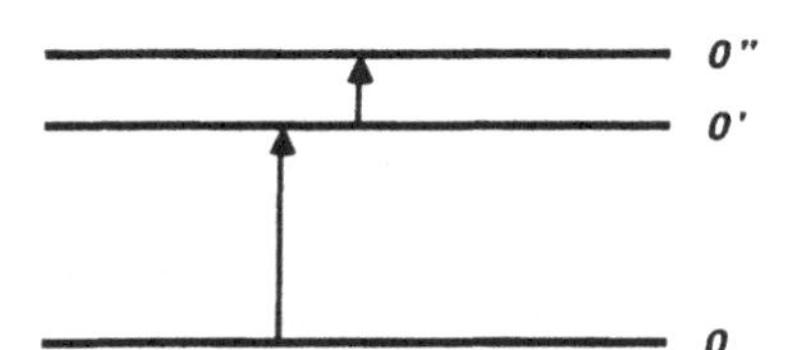

Figure 10.4. π-Electron energy levels.

the diamagnetic contribution (we are not considering spin paramagnetism). In normal metals, the quantities χ_p and χ_d almost completely cancel (there remains only a weak diamagnetic term, the so-called Landau diamagnetism). In superconductors, the paramagnetic contribution is suppressed, and χ_d dominates, resulting in the Meissner effect. When analyzing the magnetic properties of molecules, both terms must be taken into account. However, for π systems in the superconducting state, the quantity χ_p turns out to be very small (even in the absence of circular symmetry) and the molecule displays anomalous diamagnetism, analogous to the usual superconductors. Therefore, in these molecules one can observe a combination of metallic properties (delocalization of π electrons), and an anomalous diamagnetism and the presence of energy gaps. This combination is due to pair correlation of the superconducting type.

As was mentioned above, the pairing is due to the polarization of σ-core. In other words, it might be caused by a strong electron-vibrational coupling, analogous to the case described above [see Eq. (6.13)]. Large molecules (with $\gtrsim 40$ π-electrons) are close to instability. It means a strong non-adiabaticity of molecules, which is a manifestations of a strong coupling. Another channel for the pairing is connected with σ-π virtual electron transitions.

The superconducting state of molecules is apparent in a number of other properties (appearance of new energy levels lying inside the energy gap, peculiar behavior of molecules in an external electric field, etc.), but we shall not discuss them here.

Complex molecules are part of biologically active compounds. It is well known to biologists that any living matter necessarily contains conjugated systems, and consequently mobile electrons. Biologically active matter is characterized by high resistance to external irritants (recall, for example, how the human body keeps its temperature almost constant under strong external temperature variations) and by ability to transmit stimuli, that is, long-range coupling. It is quite possible that the extraordinary stability is due to the cohesion which exists in a superconducting electron system and the long-range correlation is related to the presence of bound electron pairs. Indeed, electron pairs in usual superconductors have remarkably large dimensions ($\sim 10^{-4}$ cm; see Chapter 2). One electron reacts to changes affecting its partner which is 10^4 lattice periods away! It is a unique case of long-range correlations in inanimate matter. Quite possibly, a similar mechanism is at work in biologically active compounds.

We conclude with a quote from the book *Quantum Biochemistry* by Pullman and Pullman (1963):

> . . . all the essential biochemical substances which are related to or perform the fundamental functions of the living matter are constituted of completely or, at least, partially conjugated systems. . . . The essential fluidity of life agrees with the fluidity of the electronic cloud in conjugated molecules. Such systems may thus be considered as both the cradle and the main backbone of life.

11

MEASUREMENT TECHNIQUES

Since the discovery of superconductivity, there have been very many measurements made on superconducting materials. These measurements were part of the attempts to understand the nature of the superconducting state and utilize the unique properties that have been discovered. When a researcher is working with new superconducting materials, it is essential to be able to measure at least a few critical parameters that define the fundamental phase boundaries that were discussed in the introductory chapters. These fundamental measurements involve the critical temperature, the critical current density, and the critical magnetic field. These properties will enable any researcher to evaluate the potential of these new materials for many of the applications described in this book.

CRITICAL TEMPERATURE

The critical temperature, T_c, defines the phase boundary between the normal conducting and the superconducting state, and at this temperature the resistance vanishes and the material becomes a perfect diamagnet. These two unique characteristics of the superconducting state immediately suggest two methods for determining the transition temperature, one based on resistance, the other based on magnetic susceptibility. Although a resistive measurement may be satisfactory for materials that are known to be homogeneous and superconducting, a susceptibility measurement can often detect the presence of a superconducting island in a nonsuperconducting sea. Both techniques will be discussed, but the preferred method is the susceptibility technique, as will become clear from the discussions below.

The resistance measurement requires the sample to be contacted with four leads as shown in Fig. 11.1. The two outermost leads are for the current and the

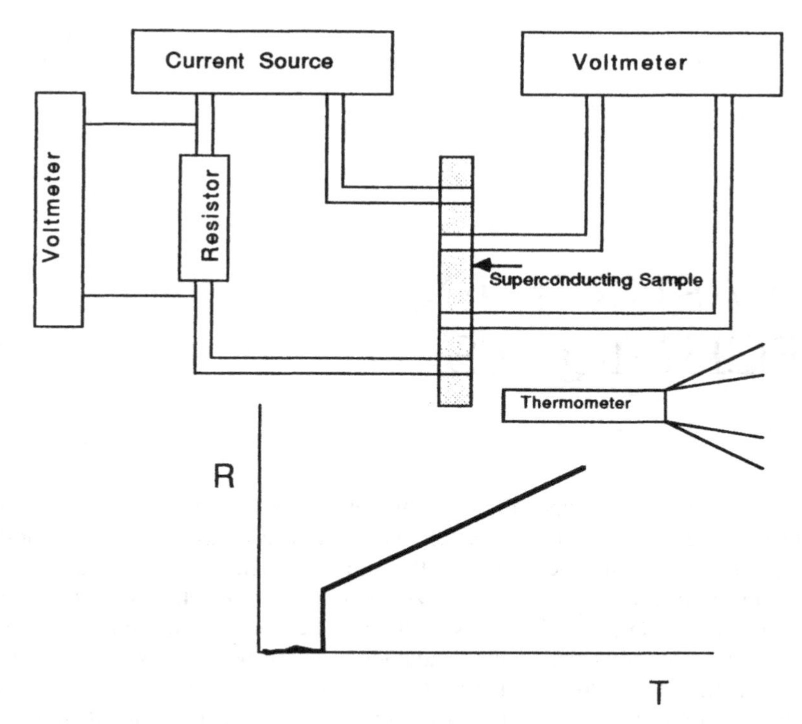

Figure 11.1 Schematic of standard four-lead resistance measurement. The lower curve is a typical transition.

two inner leads are for the voltage. It is important to have low-resistivity contacts to the sample that are ohmic. (Ohmic contacts are metallic and do not rectify the current.) The simple circuit shown in Fig. 11.1 is connected. The resistor in series with the sample is included to monitor the current through the sample. This is done to verify that the current does not change during the course of the measurement. A small current is passed through the sample, and the voltage across the sample as well as the series resistor is monitored as a function of temperature. At the transition, the voltage will decrease into the noise level of the voltmeter and the current will remain constant. For a symmetric transition, the transition temperature is usually defined as the temperature at which the resistance has fallen to one-half of its normal conducting value just above the start of the transition.

There are several points to consider:

1. The resistive transition is not a bulk measurement in the sense that if only a single superconducting filament were to connect the voltage leads, then this measurement would not reflect the majority of the sample.

2. If somehow the sample became insulating, then the voltage would also drop to zero, but the current would follow and one could immediately tell the difference between a superconductor and an insulator.

3. A crack could develop on the sample that would have the effect of isolating the current and voltage leads (see Fig. 11.2) in such a way as to cause the voltage to go to zero but the current to remain constant. The remedy for this last effect is to permute the leads and put current in leads 1 and 3 and measure the voltage between leads 2 and 4. If the same result is obtained as when putting current in leads 1 and 4 and measuring the voltage across leads 2 and 3, the sample has at least a superconducting filament.

4. The measurement never actually determines a true zero resistance but is limited by the sensitivity of the voltmeter. In fact, zero resistance has never been measured, but measurements of persistent currents in solenoids have shown experimentally that superconductors have resistivities of less than $10^{-23}\Omega$-cm. For a resistance measurement to be conclusive, the measured resistivity in the "superconducting" state must be much less than that of copper at a comparable temperature.

The ac susceptibility method for determining the transition temperature is also quite simple. The sample is encased inside a several-hundred-turn coil to which it is very closely coupled, and both the sample and the coil are thermally anchored to the platform whose temperature will be varied. A low-frequency ac signal is applied to the coil, and the ac voltage (frequencies from 1 Hz to 1 kHz and amplitide corresponding to magnetic fields of about 1 G) across the coil is monitored by a phase-sensitive detector (lock-in amplifier). A second coil is often wound in a direction opposite to the first and connected in series with the original coil but does not have a sample in it. Thus, the second coil is used to make the measurement differential so that any background temperature dependencies in the

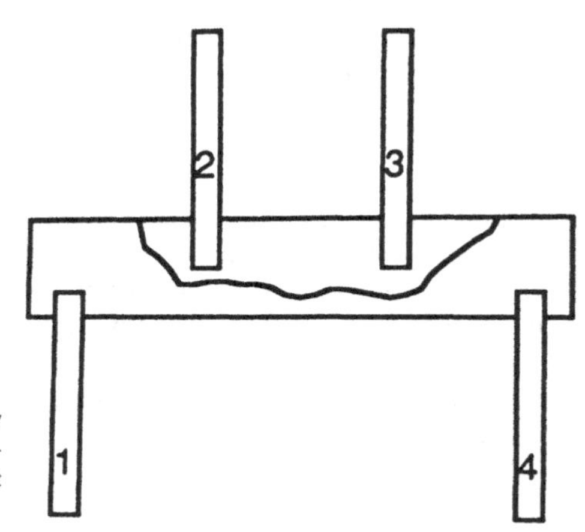

Figure 11.2. A crack that opened up at low temperature would simulate superconductivity if further experimental checks were not performed.

coil itself are canceled out. When the sample is normal, the ac magnetic field produced by the coil extends throughout the sample. [At the low frequencies involved in these measurements (<1000 Hz), the skin depth is typically much larger than the sample dimensions]. The eddy current losses in the sample contribute to the resistive part of the coil's effective impedance, but the inductive part of its impedance is only slightly changed from the value with the sample removed. As the sample becomes superconducting, the ac fields are excluded from the interior by superconducting shielding currents that flow in opposition to the applied field. This drastic change in the field profile has a very significant effect on the inductance of the sample coil, which shows up as a very significant change in the out-of-phase component of the voltage across the coil (relative to the phase of the current measured across the resistor in series with the coil). The eddy current losses are also reduced for two reasons: (1) the fields are now contained within a penetration depth of the sample surface, and (2) these currents are much less lossy than the normal eddy current shielding currents. This change is reflected in a reduction of the amplitude of the resistive or in-phase component of the voltage across the coil. In most instances, the inductive changes are larger than the resistive effects so that monitoring the out-of-phase component is ideal for detecting the superconducting transition. Often it is important to monitor both the in-phase and the out-of-phase part of the signal. In this case, a mutual inductance bridge increases the sensitivity of the measurement of the in-phase component. Such a system is schematically illustrated in Fig. 11.3.

Even with this method, there are several potential pitfalls. The inductive signal amplitude may not necessarily be connected with the volume of superconducting material contained within the sample. If the superconducting material forms a shell around the sample (like an M & M), then the signal would be the same as if the sample were completely superconducting. A test for this would be to divide the sample into increasingly smaller pieces and test each subsample. The inductive signal should scale with the sample volume. This procedure done several times would be quite convincing (but not absolute) proof that the original material was completely superconducting. Another pitfall is to connect an extremely small signal with a minute fraction of superconducting material in an otherwise nonsuperconducting matrix. A region of the sample undergoing an insulator-to-metal transition can, in some instances, mimic a superconducting transition. In this case, the signal should be frequency dependent, vanishing as the frequency approaches zero. A superconducting transition will remain down to zero frequency (a dc susceptibility measurement can, in fact, be made using a SQUID susceptometer). Often one wishes to prove that a sample is indeed a new superconductor, in this case the most convincing experiments would be measurements of both the resistance and the susceptibility since they probe the two most important aspects of a superconductor, the zero resistance and the zero magnetic induction.

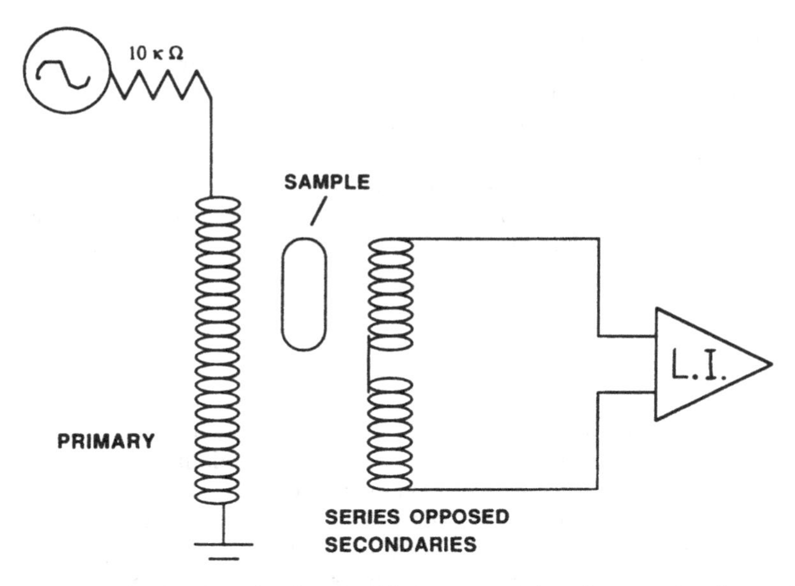

Figure 11.3. A simple schematic of a mutual inductance technique for measuring the transition temperature. L.I.: lock-in amplifier. This technique is commonly referred to as ac susceptibility.

Of course, there are many other methods to determine T_c, including high-frequency susceptibility, heat capacity, and thermal conductivity measurements, which take advantage of the remarkable difference in transport behavior and entropy of the superconducting state relative to the normal state. These methods are more complicated and will not be described here.

CRITICAL CURRENT MEASUREMENTS

Determining the maximum current that a superconducting wire or film can carry is more complicated than one would imagine because the critical current density is not an intrinsic property of the superconducting state. To understand the complexity of this problem, one must understand the behavior of a superconductor in the mixed state since most practical superconductors are Type II with rather large values of κ. Even though there are many reports of critical current measurements made with zero applied field, these will be excluded from this book. The interpretation of the zero-field results depends on detailed knowledge of the geometry of the superconductor to be sure that the critical current is not being limited by the current-induced fields which are dramatically affected by the sample geometry. If the critical current is mainly due to the entry of magnetic flux into the sample at H_{c1}, then in this case the critical current is inversely proportional to the radius of the sample and represents the current at which vortices first enter the sample. This effect always occurs in Type I superconductors, where the critical current occurs when the surface field is equal to H_c. This is called Silsbee's rule. Similar effects in Type II superconductors have been observed for the cuprate high-Tc materials.

In the presence of a magnetic field exceeding H_{c1}, vortices are always present in the superconductor, forming a well-defined triangular lattice structure whose lattice parameter a_0 is given by the expression

$$a_0 = \left(\frac{\phi_0}{B}\right)^{1/2}$$

where ϕ_0 is the flux quantum, and B is the magnetic induction.

If a current is applied to the superconductor containing this vortex array, there is a Lorentz force $\mathbf{j} \times \mathbf{B}$ acting on the vortices, and they will move across the sample if they are not pinned to defects in the superconducting sample. The motion of vortices is a dissipative process so that this motion gives rise to a finite resistance. If all the vortices move, then the resistance is just proportional to the magnetic field. In fact, at low temperatures and for fields much smaller than H_{c2}, the ratio of this "flux flow" resistance to the normal resistance of the sample is equal to the ratio of the applied field to H_{c2}^*, where H_{c2}^* would be the value of the upper critical field in the absence of paramagnetic limiting.

Most superconducting wires are specially fabricated to have many pinning sites which act to effectively pin the vortex lattice and significantly reduce the resistance to a value many orders of magnitude below what can be easily measured in the laboratory. The resistance however is not absolute zero anymore but remains finite due to a phenomenon known as "flux creep." This phenomenon was observed in the early 1960s by Kim *et al.* and explained by Anderson (see in Parks, 1969). The explanation simply involves the thermal excitation of flux bundles out of the potential well they occupy when they are in the lattice structure. Once they are excited out of their potential well, they are driven across the sample by the Lorentz force and there is dissipation. This is a thermally activated process that depends strongly on the complex interaction of vortices with other vortices and with the large number of pinning sites in the material. The resistivity would therefore increase exponentially with some activation energy U_0 divided by k_B. For the technologically significant conductors made from NbTi and Nb_3Sn, this finite resistivity is so small as to be negligible. It only would affect the very long-term decay of a persistent current set up in a superconducting solenoid. "Flux creep" is very significant for the cuprates and is one of the factors limiting their usefulness at the present time.

Now that we understand that a superconductor in the mixed state has properties that depend on the microstructure and pinning effectiveness, we realize that the definition of critical current under these conditions does not signal the onset of finite resistivity but rather signals the crossing of some arbitrary threshold that makes the material no longer functional in the sense of "negligible" joule losses.

Thus, we need a criterion to define the critical current, and this criterion can either be a resistivity criterion or an electric field criterion. These criteria are simply related by Ohm's law if we write it in the following way:

$$J = E/\rho$$

Thus, if we choose an electric field criterion of 1 mV/cm, then the value of current per unit of cross-sectional area through the superconductor that causes this value of the electric field between the voltage leads is called the critical current density J_c. However, according to Ohm's law, the field value divided by the critical current density is equal to the value of the resistivity of the sample when it crossed this threshold. For example, if the current density that gave the 1-mV/cm electric field was 100 A/cm^2, then the resistivity of the sample at this point was 10^{-8} Ω-cm, which is comparable to that of copper at 4 K. Obviously in this case, more information is required to determine if the material really has negligible losses at currents less than the current that was necessary to cross the electric field threshold. What is required is a complete current–voltage (I–V) characteristic. This characteristic is then fitted to a power law in the region around the threshold so that we model this characteristic by the equation

$$V = I^n$$

If n is large (>20), then we have a truly representative critical current that is not very dependent on our electric field criterion, but if n is between order 1 and 5, then we are not making a very useful measurement of critical current and the indication is that there are significant joule losses at much lower currents than are required to exceed this threshold. Of course, we could choose a much smaller field criterion, but, unless we have very long samples, this presents extreme difficulties in making the measurements.

As a guide, the ASTM has suggested a 10^{-12} ohm-cm criterion for resistivity, which is about four orders of magnitude lower than the resistivity of cryogenic copper. This criterion was developed for the multifilamentary conductors that are available in long lengths, and typically the critical current measurements are done on a rather long (>10 cm) sample. The exponent n for the I–V characteristic is almost always larger than 20, and this behavior is typically due to the manner in which the supercurrent redistributes itself after some filaments are driven into the normal state.

One practical note of caution should be stated—namely, the contacts to the sample must have very low contact resistivity. This will prevent joule heating in the contacts from destroying the superconductivity in the sample rather than the current in the sample. This problem has been significant in the attempts to make measurements on the cuprate superconductors.

Again in summary, in order to make meaningful critical current measurements, the applied magnetic field must be larger than the lower critical field H_{c1} in order to remove the possibility of the measurements being dominated by a Silsbee's rule type geometrical effect. The electric field criterion should be chosen so that the resistivity below the threshold is in a useful regime, that is, much less than that of copper at a comparable temperature. Finally, an I–V characteristic should be recorded in the region of the threshold to determine the threshold sensitivity of the critical current density.

Estimation of the Critical Current Density with the Use of the Critical State Model

Often one needs to estimate the critical current density of a sample on which it is just not practical to perform a direct transport measurement, for example, a small single crystal. In this case, the critical current density can be estimated from a magnetization measurement by utilizing the hysteresis in the magnetization curves. The model that is typically used is the critical state model. This model assumes that the critical current is flowing in the sample to prevent the penetration of flux into the sample when the magnetic field is increased beyond H_{c1}. As an example, consider what the magnetization would look like if there were no flux pinning and therefore a negligible critical current density, as defined by any criterion, above H_{c1}. As we increase the field from zero, the magnetization of the sample [assume that the sample is in the shape of a cigar with the field along its axis so that demagnetization (geometrical) effects are negligible] will increase as $-4\pi H$, illustrating the perfect diamagnetism of a superconductor. Above H_{c1}, the flux will enter the sample *uniformly* as vortices and the magnetization will significantly decrease. It will continue to decrease as the field is increased until H_{c2} is reached, at which point it will be zero, the field distribution in the superconductor will be uniform, and the sample will be completely normal. This behavior is illustrated in Fig. 11.4. If the field is now decreased, this curve will reproduce itself in reverse and there will be no hysteresis. The flux moves in and out of the superconductor without any barriers.

If there are significant pinning centers, then the magnetization curve will look entirely different. Above H_{c1}, instead of flux moving in freely, shielding currents up to the critical current will flow to prevent the equilibrium number of vortices from entering the superconductor. The critical state model assumes that the critical current flows uniformly throughout the superconductor (critical state) to shield the applied field. This implies that the field profile is a linearly decreas-

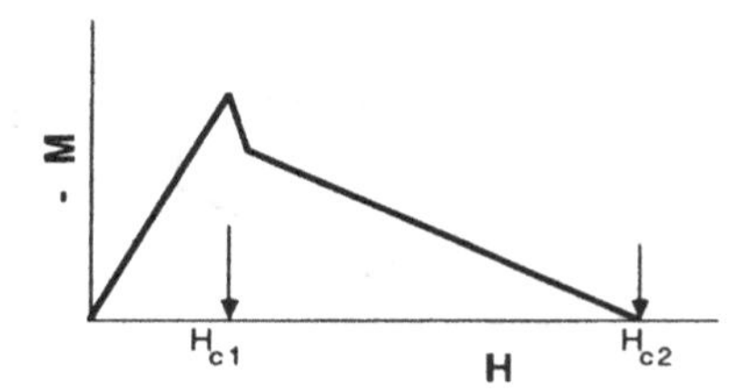

Figure 11.4. Reversible magnetization curve of a Type II superconductor.

ing function of distance from the edge to the inside of the sample; that is, dB/dx is a constant proportional to J_c. This behavior is illustrated in Fig. 11.5a. The effect on the magnetization curve is illustrated in Fig. 11.5b, where the magnetization above H_{c1} is now much larger than in the reversible equilibrium case shown in Fig. 11.4. As the field is increased, the field profile changes as is illustrated in Fig. 11.5a. Of course, as the field approaches H_{c2}, the critical current decreases and the field profile eventually become flat and equal to the applied field at H_{c2}. At this point, the field permeates the entire sample. Now, as the field is decreased, the critical current now flows in the opposite direction to prevent the flux from leaving the sample because again it is pinned to defects and irregularities. Now the magnetization becomes positive because the internal field is actually larger than the applied field due to the excess flux trapped inside. This is also illustrated in Fig. 11.5b. The field profile in the decreasing field case is shown in Fig. 11.5c. As can be seen from Fig. 11.5b, there is considerable hysteresis in the magnetization curve, and, in fact, the hysteresis continues all the way to zero field since there is actually trapped flux even when the applied field is zero. It can be shown that the critical current density at any field value is directly related to the hysteresis in the magnetization curve by the following expression for a flat plate sample with the field perpendicular to the plane:

$$J_C(H) = 30 \left(\frac{M^+ - M^-}{2} \right) \bigg/ \frac{ab}{2}$$

where M^+ and M^- are the increasing and decreasing field magnetization density, respectively, in emu/cm^3, a and b are the lateral dimensions of the sample, in cm, and J_c has units of A/cm^2.

This method usually gives results that are comparable to those obtained by the direct transport method described above. Note that these measurements can also be affected by flux creep since the magnetization can be time dependent. If one were to sit above H_{c1} on the magnetization curve of Fig. 11.5b for a very long time, the magnetization would slowly decrease toward some equilibrium value, reflecting the decay of the current due to the same flux creep that gives rise to a finite resistivity in a direct transport measurement.

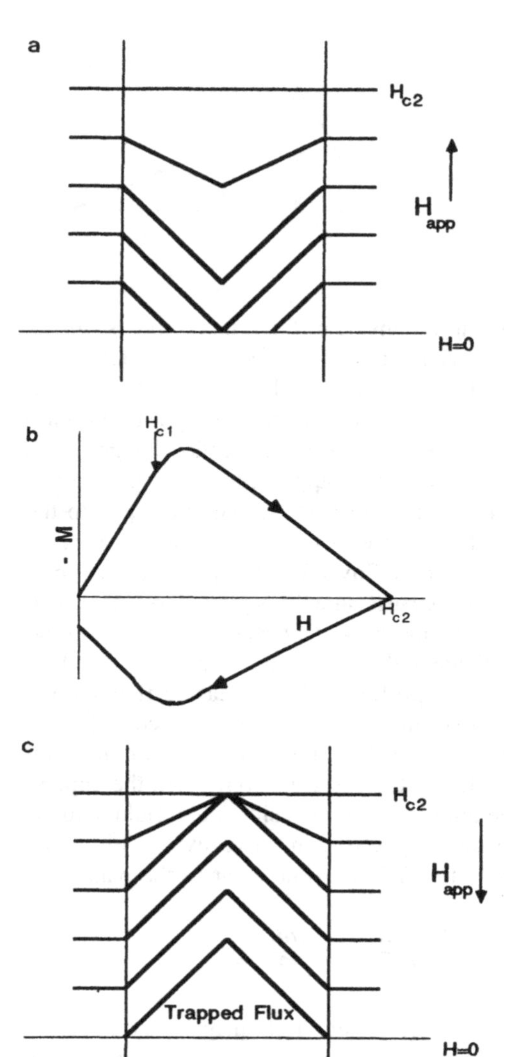

Figure 11.5. (a) The magnetic field profile inside a superconductor in the critical state model (in increasing field); (b) the magnetization curve with pinning; (c) the magnetic field profile in decreasing field.

CRITICAL MAGNETIC FIELD MEASUREMENTS

The last of the three important parameters for determining the usefulness of a new superconductor is the value of the upper critical field H_{c2}. There are several ways of determining this quantity, and two of them will be described below. The first one has been used rather extensively for conventional (noncuprate) superconductors and is based on a resistance measurement. The second, based on magnetization, has become the method of choice for the cuprates because of difficulties in interpreting resistive measurements for these unusual superconductors.

Resistive Method

The resistive method is based on measuring the resistance of the superconductor as a function of temperature at constant magnetic field. The method employed for the resistance measurement is analogous to that described above for determining the resistive transition temperature. In this case, the transition temperature is defined by the midpoint of the transition curve as a function of the applied magnetic field. An example of such a series of curves is shown in Fig. 11.6. The resulting plot of critical temperature versus magnetic field or, equivalently, critical field versus temperature can usually be fit to the curve calculated by Werthamer, Helfand, and Hohenberg (WHH) (see in Parks, 1969) for various paramagnetic and spin–orbit coupling parameters, an example of which is plotted in Fig. 11.7. Comparing the data to this theoretical curve will allow the paramagnetic and spin–orbit parameters to be extracted. If the critical magnetic field is very large so that only measurements near T_c can be made, then the simple expression $H_{c2}(0)^* = 0.69(dH_{c2}/dT)_{Tc}T_c$ can be used to obtain the critical field in the absence of paramagnetic limiting.

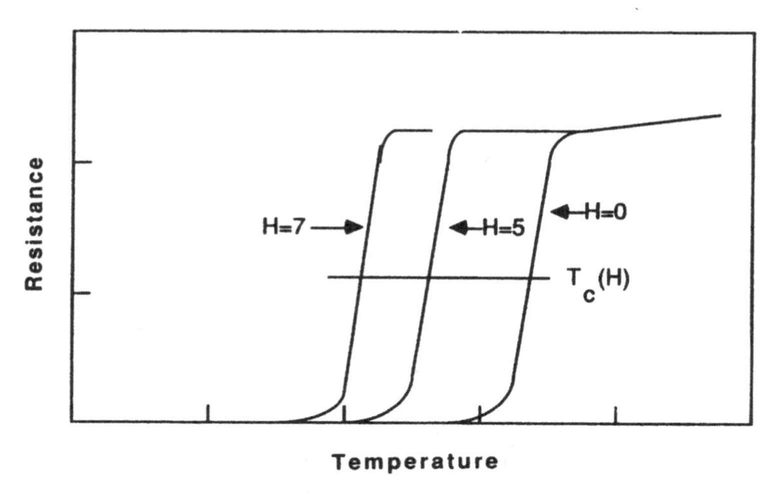

Figure 11.6. Resistance versus temperature for various values of magnetic field. The midpoint of the transition defines $T_c(H)$.

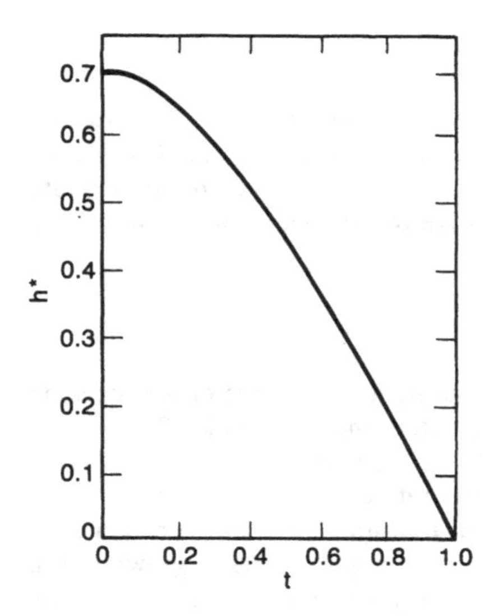

Figure 11.7. Reduced critical field $h^* = H_{c2}(t)(dH_{c2}/dt)_{t=1}$, where $t = T/T_c$, for no spin–orbit coupling and no paramagnetic limiting.

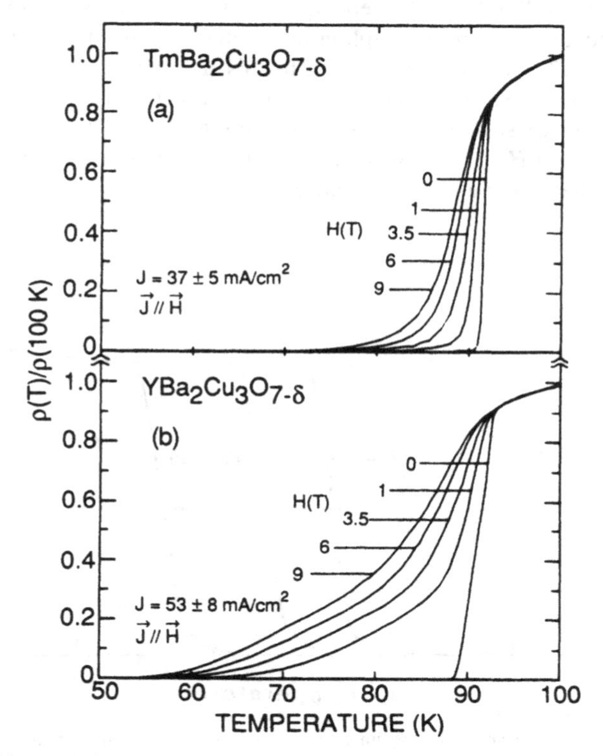

Figure 11.8. Normalized resistivity ρ versus temperature for $TmBa_2Cu_3O_{7-x}$ and $YBa_2Cu_3O_{7-x}$ in various applied magnetic fields between 0 and 9 T (see Maple in Kresin and Wolf, 1987).

Magnetization Method

The cuprate superconductors are a prime example of superconductors for which the resistive method does not work. Even for the very low currents that are utilized in the resistive method, the high transition temperatures, the low density of pinning sites, and the small pinning potential barrier due to the small value of the coherence length magnify the effects of flux creep and flux motion. Thus, the resistive transition becomes very broad and asymmetrical in a magnetic field (see Fig. 11.8). The midpoint of the transition is no longer associated with the upper critical field.

The magnetization measurement relies on the fact that $H_{c2}(T)$ is defined as the field at which the diamagnetic moment vanishes at the temperature T. The measurement is done at constant field, and the magnetization is measured as a function of temperature as the temperature is decreased from well above the transition temperature. This measurement is particularly sensitive for samples with reversible magnetization curves near H_{c2}. At relatively high magnetic fields, the magnetization in the normal state is constant and starts to decrease linearly as soon as superconductivity appears in the sample [below $T_c(H)$]. The intersection of the linear magnetization and the normal-state magnetization gives $T_c(H)$. This can then be inverted to determine $H_{c2}(T)$. Comparing the values obtained by this method with resistive transitions, there is no clear connection with any feature of the resistance curve and H_{c2} for the cuprate superconductors.

12

APPLICATIONS OF SUPERCONDUCTIVITY

The unique properties of superconductors that we have discussed earlier are the basis of very many applications that will be described in this and the next chapter of this book.

The applications will be divided into those which are basically small scale, categorized here as electronic applications, and the large-scale applications involving the conduction of large currents, which will be discussed as power applications.

SUPERCONDUCTING ELECTRONICS

There are very many electronic applications of superconductors, and it would be nearly impossible to cover them all in any great detail. In this section, some of the more highly developed electronic devices will be discussed while the more esoteric ones will just be mentioned.

Josephson Junctions

A large number of applications of superconductors in electronic devices are based on the properties of Josephson junctions. These junctions are formed by two superconducting films that are separated by a very thin insulating layer so that they are only weakly coupled and therefore their transport properties can be described by the Josephson equations and both the ac and dc Josephson effects can be observed. The current–voltage characteristic of such a junction is shown in Fig. 12.1. Notice that this I–V characteristic has two important features: the zero-voltage supercurrent and the semiconductor-like finite voltage highly nonlinear characteristic. The zero-voltage current is the Josephson supercurrent, and

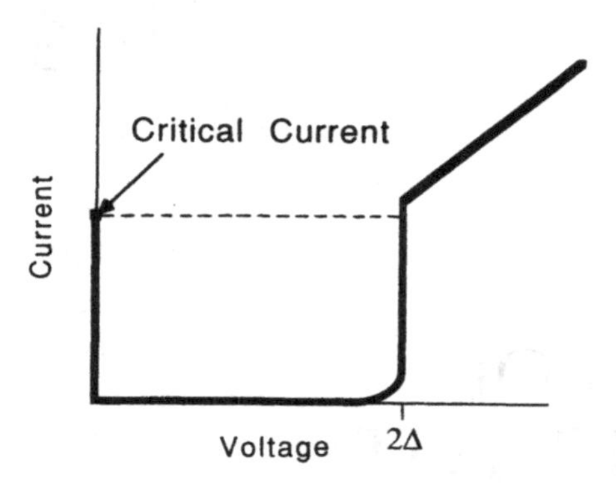

Figure 12.1. The current–voltage characteristic of a Josephson junction.

the maximum value of this current is the Josephson critical current. Using the Josephson equations, it can be readily shown that this critical current is very strongly affected by small magnetic fields and follows the following equation:

$$I_C(\phi) = I_C(0) \left| \frac{\sin(\pi\phi/\phi_0)}{\pi\phi/\phi_0} \right|$$

where ϕ is the applied flux (B·A) and ϕ_0 is the flux quantum ($h/2e$). Thus, the critical current is modulated by the field in an analogous manner to the Fraunhofer pattern for optical diffraction. The critical current is zero whenever an integral number of flux quanta are contained within the junction. Since the flux quantum is very small, this effect can be utilized to build a very sensitive magnetometer, as will be described below.

If the critical current is exceeded, the junction switches into the voltage state, and the voltage that appears is approximately the voltage corresponding to the sum of the two energy gaps of the superconducting films on either side of the junction. This "switching" takes place extremely rapidly, and the switching time in most real junctions is limited by the junction capacitance. The intrinsic superconducting time is given by the ratio of $\hbar$ (Planck's constant divided by 2π) and twice the energy gap of the superconductor and is typically less than a picosecond. If the current is now reduced, the voltage does not follow the same curve as for increasing current but nearly follows the semiconductor-like curve so that there is typically a large hysteresis between increasing and decreasing current. This switching and hysteretic nature of the I–V characteristic is the basis for the utilization of Josephson junctions as both logic and memory elements for digital applications. If a sufficient magnetic field is applied to the junction or if the coupling of the two superconducting films is very weak, then the Josephson supercurrent portion of the I–V characteristic disappears and the semiconductor-like behavior remains. There are some important distinctions, however, between this characteristic and that of a typical semiconductor. The value of the voltage corresponding to the knee in the I–V curve is in the millivolt rather than the volt range because the energy gap in a superconductor is approximately a thousand times smaller than that in a semiconductor. For very high quality junctions, this knee can be extremely sharp. Mixers and video-type detectors can be readily made utilizing this highly nonlinear curve.

It is often important to suppress the hysteresis that is observed in the I–V characteristic of a high-quality junction. The technique that is now universally used is to deposit a resistor across the junction with a value slightly less than the normal-state resistance of the junction. This shunt does not allow the current to drop below the critical current when the voltage is reduced to below the sum-gap voltage on the decreasing-current part of the I–V characteristic. This type of junction is usually called a resistively shunted junction (RSJ).

Whenever a Josephson junction is biased into the finite-voltage state, there is an ac supercurrent that is generated in addition to the dc current that is biasing the junction. The frequency of this current is directly proportional to the voltage and is given by the second Josephson equation:

$$V = \hbar\omega/2e$$

where V is the voltage across the junction, and ω is the frequency of the ac Josephson supercurrent that flows across the junction due to the time evolution of the phase of the superconducting wave function. This relation is extremely precise and, in fact, forms the basis of the currently accepted voltage standard (see below for details). The presence of this ac current is the basis for several interesting devices which will be described below.

Typically, Josephson junctions are fabricated by more or less conventional techniques that have been perfected by the semiconductor industry. Silicon is the preferred substrate, and the superconducting film, the insulating layer, and the counter electrode (the upper superconducting film) are usually prepared as a trilayer. The junction areas are defined by photoresist and photo- or electron lithography. Anodization, ion beam, or chemical methods are used to remove the upper two layers except where the junctions are to remain. Insulating layers, resistors, capacitors, and wiring levels are then evaporated, forming the desired circuit incorporating the Josephson junctions.

SQUIDS

SQUID is the acronym for a *s*uperconducting *qu*antum *i*nterference *d*evice, which is a simple device that contains one or two Josephson junctions. As was described above, the critical current of a Josephson junction is very sensitive to the magnetic flux that threads the junction and, in fact is modulated with the

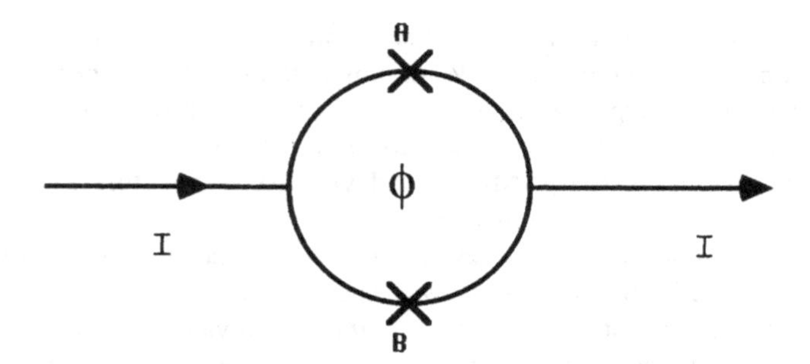

Figure 12.2. Schematic of a dc SQUID.

period of a flux quantum (2×10^{-7} G-cm^2). If two Josephson junctions with resistive shunts to eliminate the hysteresis in the I–V characteristic are placed in parallel and the magnetic vector potential integrated around the loop formed by the junctions and the connections (see Fig. 12.2), then the resulting equation for the critical current as a function of flux is given as

$$I_c(\phi) = I_c(0)|\cos(\pi\phi/\phi_0)|$$

where ϕ is the total flux enclosed in the loop, and $I_c(0)$ is the zero-flux critical current. This equation is identical to the equation describing the quantum interference of light diffracted by two slits. An example of such a modulation is shown in Fig. 12.3. This analogy of the response of this two-junction "interferometer" is the origin of the name SQUID. The important difference between this modulation and the modulation of a single junction is that the zeros of the modulation occur when the flux through the loop changes by an integral number of flux quanta, but in this case the appropriate area is not that of the junction, but that of the loop containing the junctions. Thus, the magnetic field sensitivity of this device is quite remarkable. For example, for a 1-cm loop area, the zeros of the modulation would occur every 0.2 μG. In fact the actual resolution of such a device can be much better than a single flux quantum. The normal approach to making a practical device is to apply a dc bias current that is close to the critical current at zero flux and an ac flux modulation to the SQUID via a small coil coupled to the SQUID loop, about one flux quantum in amplitude and at a sufficient frequency to be about an order of magnitude above the maximum frequency that is going to be measured. This modulation produces a modulation of the critical current that can be detected as a voltage modulation across the input to the SQUID. This voltage is phase-sensitively detected and fed back to the coupling coil in such a manner as to produce a minimum response; that is, the modulation produced is symmetric about the minimum critical current. Any flux change to the input coil is counteracted by a change in the feedback current, which is what is measured. This feedback current is proportional to the magnetic flux change through the SQUID loop. The resolution that can be obtained routinely is better than one part in one hundred thousandth of a flux quantum in a 1-Hz bandwidth. In practice, one does not make a very large SQUID loop but uses what is called a flux transformer instead. This device works on the principle that in a closed superconducting circuit the flux is constant and remains fixed at the value contained in the circuit when the material becomes superconducting. This flux transformer has two coils: one is inductively coupled to the SQUID loop (coupling coil), and the other is exposed to the magnetic environment that is to be measured (input coil). Typically, the SQUID itself and the part of the flux transformer that is coupled to it are very well shielded from the magnetic background by a superconducting shield which preserves the magnetic environment that exists when the material becomes superconducting. The flux inside this shield does not change as long as the material remains in the superconducting state. As the magnetic field coupled to the exposed coil of the flux transfomer changes by ΔB_i, a current is induced in the flux transformer that keeps the total flux enclosed constant. If L_i is the inductance of the input coil and L_c the inductance of the coupling coil then the current, Δ_i, produced is $(\Delta B_i \cdot A_i)/(L_i + L_c)$ where A_i is the area of the input coil. The change in field at the SQUID coupling coil is Δ_i

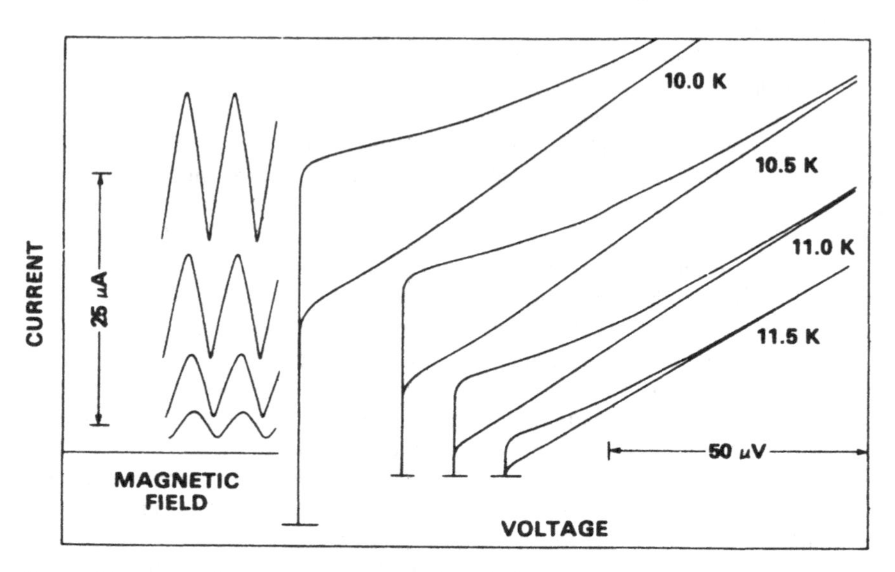

Figure 12.3. *Right:* Current–voltage curves of a SQUID, corresponding to the maximum and minimum critical currents induced by varying the magnetic flux. *Left:* Variation of the current at ~2μV with magnetic flux, for the same sequence of temperatures.

L_c/A_c. If L_c is equal to L_i then the change in field at the SQUID is $(\Delta B_i A_i)/2A_c$. If A_i is much larger than $2A_c$ as is generally the case, then there can be significant field multiplication using such a transformer. Even more useful is the capability of making gradiometers. A first-derivative gradiometer can be made by dividing the input coil into two parts wound in opposition and separating them by some distance Δx. In this case, for a uniform change in magnetic flux, the currents in each of the separate coils will be equal and opposite so that the net change in current will be zero. If, however, there is a gradient, $\Delta\phi/\Delta x$, in the flux change, then the two induced currents will not exactly cancel, and the net current will be proportional to the flux or field gradient. In a similar manner, higher-order gradiometers can also be wound. This makes these devices very sensitive for local sources of magnetic flux but quite insensitive to the background Earth's field. These devices are the most sensitive magnetometers and gradiometers that have ever been built and are becoming quite useful in biomagnetism and for magnetic anomaly searches, both in prospecting and in antimine and antisubmarine warfare.

Another SQUID device can be made using a superconducting loop that contains only one resistively shunted junction. This device is called an rf SQUID. The loop containing the junction is inductively coupled to an LC circuit that is resonant in the megahertz region (rf) with a quality factor of at least 100. As the instantaneous current in the inductor increases, a current is induced in the SQUID loop that cancels the flux produced by the current in the inductor. When this shielding current reaches the critical current of the junction, the loop can no longer shield the applied flux but allows one flux quantum to enter the loop, decreasing the shielding current by $\phi/L(\text{sq})$. The shielding current is then less than the critical current, and the loop can shield again. Thus, the loop with one junction behaves like a flux turnstile, allowing flux to enter one quantum at a time whenever the critical current is reaches. The entry of flux into the loop is a dissipative process, and when this occurs every rf period, the effective quality factor of the resonant circuit is diminished. This decrease in Q can be directly observed by measuring the dc voltage across a diode detector. In a plot of this rf detected voltage versus rf drive current, when the rf drive current is sufficient to cause the SQUID loop current to reach its critical value, the slope is drastically reduced. If a low-frequency flux is also applied to the SQUID loop, shielding current also flows to cancel this flux so the total shielding current is the sum of the rf and the dc value so that the position of the break in the rf I–V characteristic is a function of the applied flux. In fact, if a slowly varying flux is applied to the SQUID loop, this break in the I–V curve is modulated up and down with a period of one flux quantum in a triangular pattern so that if the rf level is set so that the critical current is just reached in the absence of any other flux, then the detected voltage will undergo a triangular modulation with a flux period of one quantum. Again the appropriate area is the loop area, not the junction area, since the shielding currents are flowing around the loop. Again this device can be phase locked to act as a null detector, and it is the feedback current that is proportional to the applied flux (or field). A flux transformer can also be used to couple the flux into the SQUID loop with some amplification. The rf SQUID is in many ways a much simpler device than the dc SQUID since it is only inductively coupled to the outside, there are no current or voltage leads, and it requires only one junction for fabrication. However, for reasons that cannot be simply discussed here, for typical operating frequencies (20 MHz) the resolution in fractions of a flux quantum per $Hz^{1/2}$ is not as good as that of the dc SQUID. A much higher frequency bias would be required for the rf SQUID to become equal in resolution to the dc SQUID. The rf SQUIDs were the first to become commercially available, but dc SQUIDs can now also be purchased.

Detectors

Superconducting junctions can be used to detect electromagnetic radiation over a very broad range of frequencies as well as with several different modes of detection. Furthermore, because of its small thermal mass and sharp transition from the normal resisting state to the zero-resistance superconducting state, a superconducting thin film, biased at the transition edge, can be a very sensitive detector of the temperature rise due to the absorption of any radiation.

Josephson Junction Detectors

A resistively shunted Josephson junction can be quite sensitive to electromagnetic radiation. There are important differences in how these junctions respond to photons whose energy is above the superconducting energy gap compared to photons with energies below the energy gap.

For electromagnetic radiation with energies below the energy gap of the junction electrodes, the effects on the I–V characteristic are quite remarkable. The ac currents suppress the dc critical current since the currents add algebraically. However, more importantly, in the finite-voltage region, the ac Josephson supercurrent can mix with the induced ac currents and produce sum and difference harmonic currents in the junction. If the ac Josephson frequency and the applied frequency are harmonics of one another, then there is a mixing harmonic at zero frequency, that is, at dc. Thus, an additional dc current will flow at these matching frequencies. By sweeping the applied voltage, a large range of frequencies can be sampled according to the Josephson voltage-to-frequency conversion factor of 484 MHz/microvolt. Thus, there appear constant-voltage current steps at voltages that correspond to harmonics of the applied radiation. This forms the basis of a frequency-sensitive detector. The suppression of the dc critical current can also be used as the basis of a non-frequency-discriminating detector analogous to a video diode detector. The extreme precision of the Josephson frequency–voltage relation is the basis for the standard volt that is now used in many of the standards laboratories around the world. In fact, many Josephson junctions in series are used to bring the steps up to much larger voltage values for 10-GHz radiation.

For radiation that is above the gap energy [$h\nu > E(\text{gap})$], the physics of the detection mechanism is entirely different. The radiation can cause the superconducting Cooper pairs to dissociate and produce excess quasiparticles (unpaired electrons). These quasiparticles act to decrease the superconducting energy gap and hence also cause a depression of the critical current, which can be detected by a circuit biased very close to the critical current. The radiation causes a finite voltage to appear across the junction with an amplitude proportional to the intensity of the radiation over some limited range of amplitudes. This type of device has found only limited usefulness as an isolated junction but as part of an array, either fabricated or by virtue of granularity, may be quite important for IR and optical detection.

SIS Mixers

If the insulating layer between the superconducting films is thicker than 10–20 Å, the supercurrent contribution to the dc I–V characteristic becomes negligible and the semiconductor part is all that remains. The extremely nonlinear characteristics, especially at or near the sum-gap voltage, can be used very effectively to mix or detect below-gap radiation. A typical SIS I–V characteristic both in the absence and in the presence of radiation is shown in Fig. 12.4. The step that occurs below the gap voltage is due to photon-assisted tunneling where the applied voltage plus the effective voltage $V = h\nu/e$ of the radiation sum to equal the gap voltage. Just as for a regular diode mixer, two frequencies applied to a nonlinear element will generate harmonics at difference and sum frequencies

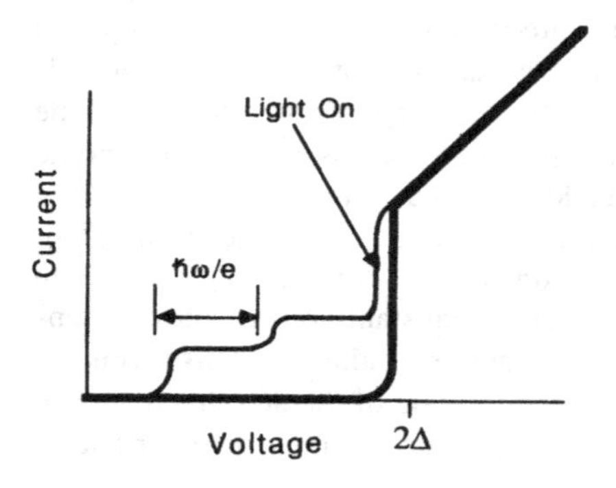

Figure 12.4. Current–voltage characteristic of an S–I–S junction with and without radiation.

and thus can both down convert and detect radiation. The sharpness of the characteristic and the fact that the noise of the best detectors is close to being in the quantum limit has made SIS mixers the detectors of choice for radio astronomy because the background radiation does not swamp these detectors owing to their limited dynamic range and very low noise temperature.

Bolometers

As was mentioned above, the rather large change of the resistance with temperature that can be obtained at the superconducting transition makes superconductors very sensitive thermometers. The sensitivity of a superconductor, or any component to which it is thermally anchored, to small amounts of heating is quite high and, when optimized by the proper isolation, is competitive with that of more conventional detectors.

Microwave Components

A very important property of superconductors is their very small value of surface resistance, R_s. Surface resistance is defined as the ratio of the joule losses and the square of the surface magnetic field. A superconductor at finite frequency has joule losses and therefore has a nonzero surface resistance. The temperature and frequency dependence of the surface resistance has the following approximate BCS form when the temperature is less than half the transition temperature:

$$R_S(T, \omega) = \frac{A\omega^2}{T} e^{-\Delta/k_B T}$$

where Δ is the energy gap of the superconductor, and A is a constant that depends on the London penetration depth in the material. This is to be compared to the surface resistance of a normal metal, which is given by the following equation:

$$R_S = \left(\frac{\mu\omega}{2\sigma}\right)^{1/2}$$

where σ is the electrical conductivity, and μ is the magnetic permeability of the conductor.

Thus, the surface resistance of a normal metal increases as the square root of frequency and resistivity whereas it goes quadratically with frequency in the superconductor. The temperature dependences are quite different as well. A comparison of the surface resistance for Nb and Cu as a function of frequency and of temperature is shown in Figs. 12.5 and 12.6, respectively. The dramatic dropoff in the surface resistance just below the transition is due to the high-

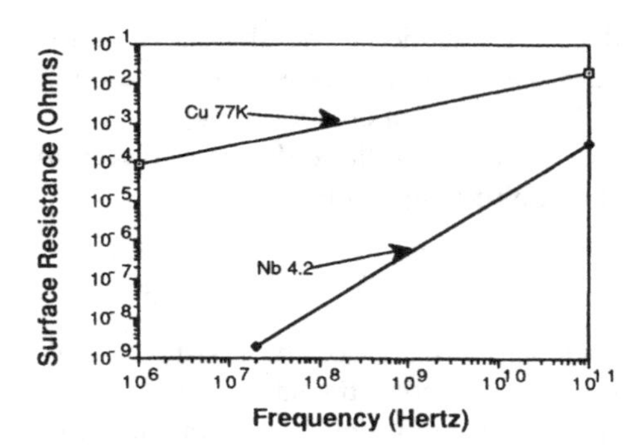

Figure 12.5. Frequency dependence of the rf surface resistance of Nb and Cu.

frequency analogue of the Meissner effect. In the normal state, the electromagnetic fields penetrate the skin depth (δ) and the losses occur in the volume defined by the surface area times δ. In the superconducting state, the losses occur within a penetration depth that is typically much smaller than the skin depth and is also not frequency dependent. The penetration depth rapidly shrinks to close to its final value within a few percent of the transition temperature. Thus, the initial drop in surface resistance reflects the decrease in the volume over which the losses occur. The exponential fall that occurs below about half of T_c represents the decay with temperature of the number of quasiparticles thermally excited into the state above the gap and able to absorb radiation and cause joule losses. The saturation at very low temperatures is not intrinsic but depends on the concentration of impurities or other nonsuperconducting regions on the surface (bad grain boundaries, oxides, etc.).

Another important point to take particular note of is the fact that the penetration depth is a very weak function of frequency up to frequencies approaching the gap frequency whereas the skin depth varies as the square root of the frequency. Since the velocity of propagation of a given frequency depends on the skin depth in a normal metal and the penetration depth in a superconductor, the dispersion

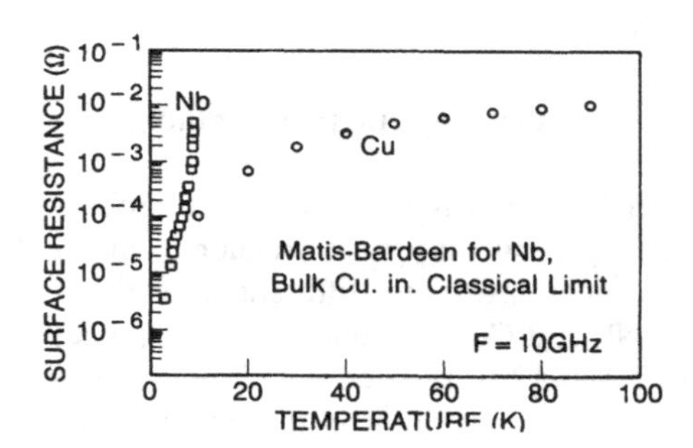

Figure 12.6. Temperature dependence of the rf surface resistance of Nb and Cu.

(dispersion is the spreading out in time of a pulse due to the propagation of the different frequencies contained in its spectrum) will be quite different in the two states. In fact, there will be almost no dispersion in the superconductor, and thus, almost independently of the particular value of the surface resistance, the pulses will propagate without spreading out in time. The combination of surface resistances that are much lower than those of the best normal conductors over a very wide range of frequencies and the nearly dispersionless propagation of signals makes superconductors especially important for microwave applications.

High-Q Elements

Most microwave electronics require high-Q elements. These elements can form part of the oscillator circuits, can be used as filter couplers, and are basically an essential part of microwave systems. Superconductors, by virtue of their extremely low (compared to normal materials) surface resistance, can be utilized as high-Q signal conditioners in many microwave applications since the Q of a resonator is given by the ratio of a geometrical factor (G) and R_s so that the microwave designer has a much greater flexibility in optimizing the signal conditioning elements of his circuit.

Signal Transmission and Delay Lines

When pulses are propagated in microwave circuits, the figures of merit are the losses per unit length and the dispersion. The larger the losses, the shorter the lines must be or the larger the amplitude of the pulses must be to ensure that the pulse amplitude does not drop below the threshold necessary for the pulse to perform its function. Superconducting transmission lines offer a distinct advantage since the losses are much smaller than for normal conductors (they are proportional to R_s). Along with the loss in amplitude, conventional conductors have considerable dispersion because of the frequency dependence of the propagation velocity. Again superconductors offer a distinct advantage because the dispersion is almost negligible for current frequencies of interest (below 150 GHz). These advantages are especially important when building delay lines which must be long to provide the appropriate signal delay (at the speed of light, this corresponds to one foot per nanosecond). Thus, building multinanosecond delay lines is a very difficult task with the use of normal conductors but can and has been easily accomplished using superconducting transmission lines.

Other High-Frequency Components

The combination of low loss and the dramatic difference between the surface resistance above and below the transition temperature makes several other devices possible. Superconducting microwave switches can be made to shift signals from one channel to another. By having several transmission lines in parallel and by switching them in and out of the superconducting state by magnetic field, dc current, or optical switching, the high-frequency signals can be directed along the superconducting paths. Another possible device is a power limiter. In such a device, the dimensions of the transmission line are adjusted so that at a given power level the rf critical current is exceeded and the line becomes much more resistive, limiting the amplitude of the signal reaching the next stage and thus protecting it from large, possibly damaging signals.

Electrically small (dimensions much less than the wavelength) antennas are also possible since the efficiency of any antenna can be shown to depend inversely on the surface resistance. Thus, very efficient but electrically small antennas can be made from superconducting materials. The antenna is no longer matched to free space but can be matched to the rest of the electronics by a superconducting matching network.

As described previously, superconducting junctions can mix, detect, and generate electromagnetic signals and can play the same role in a microwave receiver.

SUPERCONDUCTING DIGITAL ELECTRONICS

Digital applications of superconductivity are based mainly on the remarkable properties of Josephson junctions, both individually and configured in multiple-junction SQUID type structures. As was discussed previously, a hysteretic Josephson junction has two distinct voltage states for currents less than the zero-magnetic-field critical current. The junction can be switched from the nondissipative zero-voltage state to the finite-voltage state (approximately the sum-gap voltage) by the application of an additional current pulse such that the critical current of the junction is exceeded when the pulse is added to the existing current. The junction can also be switched by the application of a small magnetic field which decreases the critical current to below the existing current, causing the junction to switch to the finite-voltage state. This second method is the preferred method for switching the junction. The feature of this switch that makes it very attractive for digital applications is the possibility of producing very high speed digital circuits that dissipate very little power. In fact, the product of the speed times the power dissipation is several orders of magnitude lower than the lowest values obtained with semiconductors. The architecture of the logic and memory circuits of course has to be quite different from that for semiconductor circuits, but these details have been worked out by many groups, and digital microprocessors have been built that are about 100 times faster than the comparable semiconductor design and dissipate about 100 times less power. As an example of how such an element performs logic, consider three Josephson junctions in parallel with a common feed so that the current is divided equally between the three junctions. Next to each junction is a control line that can provide enough magnetic field to switch each of the junctions from the zero-voltage to the finite-voltage state. Each junction can carry two-thirds of the total current, but not all of the current, and still remain in the zero-voltage state. This circuit is an AND gate, because if junction 1 is switched, then junctions 2 and 3 each carry half the total current and they remain superconducting, but if both junctions 1 and 2 are switched, then junction 3 will switch and a voltage will appear across all three, signaling a coincidence or an AND between junctions 1 and 2. In contrast, switching junction 1 or junction 2 will not produce a voltage across junction 3. If the bias is increased so that half the total current can cause each junction to switch, this circuit becomes an OR gate. A memory cell can be made in several ways. The simplest is to use a hysteretic junction and to maintain the junction in the zero-voltage or finite-voltage state. Zero voltage is a "0" and finite voltage is a "1." The junction can always be reset by a negative current pulse. The disadvantage of this type of memory cell is that when the junction is in the finite-voltage state, there is dissipation and thus power is lost continuously. However, a memory cell made up of two junctions in a SQUID configuration can store information with the junctions both in the superconducting state. This memory cell consists of two junctions in parallel, with control lines for producing small magnetic fields, and readout junctions with their own control lines coupled to the memory cell junctions (see Fig. 12.7). The information is stored with circulating currents in the SQUID loop formed by the two junctions. A clockwise current produces at least one flux quantum inside the loop with the field direction up (this is a "0"), whereas a counterclockwise current produces a fluxoid with the field direction down (a "1"). These circulating currents are produced as follows. A current is passed through both junctions, from top to bottom, dividing equally between them. Then, for example, the left control line is activated, driving the left junction normal, and all the current is passed through the right junction. The control line is then deactivated simultaneously with the turning off of the bias current. Now a clockwise circulating current is trapped in the SQUID loop, and this is the "0" state. Similarly, if the right junction was switched normal and the same procedure followed, a counterclockwise current would flow and a "1" would be produced. These states can be read as follows. The readout junction cannot be driven into the finite-voltage state by either its control line or the current through the memory junction; the fields from both must add to switch the readout junction. So, for example, let us suppose that a clockwise current is flowing in the SQUID loop and that the current in the readout junction control line is from top to bottom. The magnetic fields from the two currents will nearly

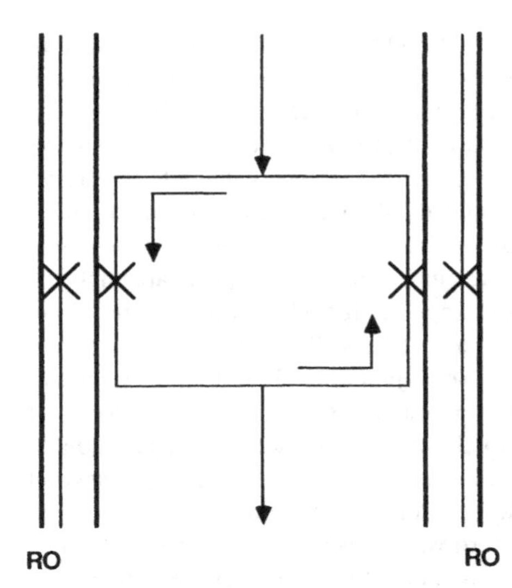

Figure 12.7. Schematic of a simple Josephson junction memory circuit. Its operation is described in detail in the text.

cancel and the junction will not switch; therefore, it reads a "0." If, however, the current flowing in the SQUID loop is counterclockwise, the fields produced by the two currents will add and the readout junction will switch to the finite-voltage state and signal a "1." Thus, although this memory cell is more complicated, it is dissipative only in the writing and reading stages, but all of the junctions are superconducting during the main storage cycles.

Of course, much more complicated architecture has evolved, and there are very good circuits for analog-to-digital converters that are very fast and have many bits of resolution. Microprocessors and memory chips have also been designed and built [see Van Duzer and Turner (1981) for more details].

POWER APPLICATIONS OF SUPERCONDUCTIVITY

With the discovery in the early 1960s of superconducting materials that have negligible resistance and can carry very large currents in very large magnetic fields at liquid He temperature (4.2 K) without becoming normal, many large-scale applications of superconductivity become possible. There are two distinct types of high-power applications: those that involve conductors for magnets or power transmission, and those that require high-power microwave cavities for accelerators or for microwave energy storage. This chapter will treat many of these applications including those that have been perfected, those that are being developed, and those that are "pie in the sky."

Conductors

The manufacture of conductors for carrying large amounts of current in high magnetic fields has been a difficult and time-consuming task, but there are presently several conductors that are quite satisfactory for building many of the systems that will be described below. The workhorse superconductor for magnets that produce fields up to 10 T is NbTi. This ductile superconducting alloy can readily be drawn or extruded so that its mechanical properties are excellent for wire fabrication. In addition, it can be coextruded with copper, and thus a matrix of fine NbTi filaments in a copper matrix can be fabricated. The structure of the superconducting wire, shown in cross section in Fig. 12.8, is complicated for many reasons relating to the operation of the superconducting magnet that will be made from it. The filaments of superconducting NbTi also have a complicated microstructure that is necessary to pin the vortices that are always present in the mixed state. It is a fine-grained microstructure with a well-dispersed second phase which pins the vortex lattice. The large number of fine filaments that are in fact twisted (this cannot be seen from the figure) reduce the inductance of the wire itself and thus reduce the eddy current losses that occur when the current is changed rapidly. The large amount of copper matrix acts to stabilize the conductor against transients which may cause some part of the conductor to go into the normal state. This stabilization spreads the heat rapidly so that the tendency is for the wire to regain the superconducting state. Thus, the wire is in stable, rather

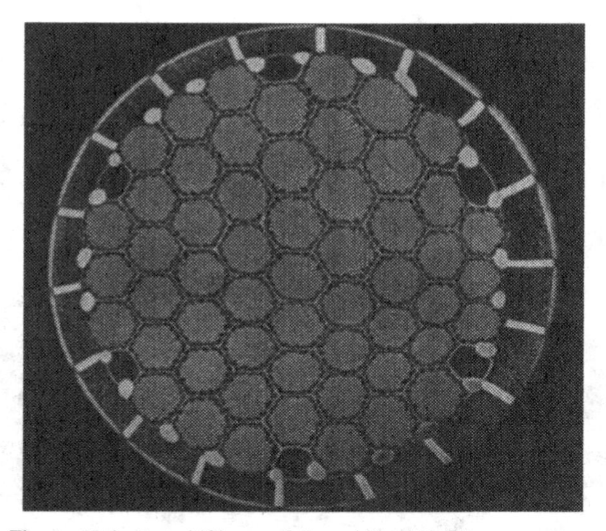

Figure 12.8. A multifilament fully stabilized NbTi superconductor.

than unstable, equilibrium with respect to small fluctuations into the normal state.

The cross section of state of the art Nb_3Sn or V_3Ga multifilamentary conductors is quite similar. These wires are very brittle and cannot be wound into small magnets unless they are reacted after winding (see section on A15 materials in Chapter 9). These wires can produce magnets with fields exceeding 20 T.

Magnets for High-Energy Physics

One of the first users of superconducting magnets were the high-energy physicists. Superconducting magnets could provide large volumes of high fields that were just not possible to obtain with conventional conductors. The bubble chamber magnets were a prime example of this large-scale use of superconductivity, and an early example of such a magnet is shown in Fig. 12.9. Soon after, it was realized that superconducting bending magnets could be used to shrink the size or increase the energy of accelerators by virtue of the higher magnetic fields they could produce and the smaller radius of curvature they could provide for a given energy. Such bending magnets at the Fermi National Laboratory (see Fig. 12.10) have made many high-energy physics experiments possible.

Figure 12.9. A very early bubble chamber magnet built at the Argonne National Laboratory.

Figure 12.10. Photograph showing the superconducting magnets installed under the existing copper electromagnets in the tunnel of the synchrotron at the Fermi National Laboratory.

Energy Storage

The energy that is stored in a magnetic field is quite considerable. In fact, it is quite easy to show that the energy stored in a superconducting solenoid is given approximately by the following equation:

$$\text{Energy stored} = B^2 \times \text{volume}/\mu_0$$

Therefore, a very large solenoid can store a considerable amount of energy, and in fact a large superconducting magnet was used by the Bonneville Power Commission to act as a load leveler to smooth out large variations in the power load. Much larger magnets can actually be used on a diurnal basis to store energy at night when the load is low and deliver it to the power grid during the day when the load is high. This type of magnet can also store the energies required for very large laser or other pulsed power applications.

Motors and Generators

Of course, magnets play a crucial role in power and motion generation so it is not surprising that superconducting magnets can play an important role here as well. There have been many different motor and generator designs proposed (and some built) using superconducting magnets, but one design seems to offer significant advantages and it will be described here as an example. The homopolar direct current motor and generator design is very applicable to superconducting magnets since the field is produced by a static superconducting magnet and the rotor is a conducting disk or cylinder (see Figs. 12.11 and 12.12). Small motors and generators based on this design have been successfully built and operated. Ship propulsion is one of the major areas in which this technology will have an impact.

Magnets for Magnetic Resonance Imaging

Magnetic resonance imaging (MRI) is one of the medical wonders of the last decade. This technique, which probes the nuclear magnetic resonance

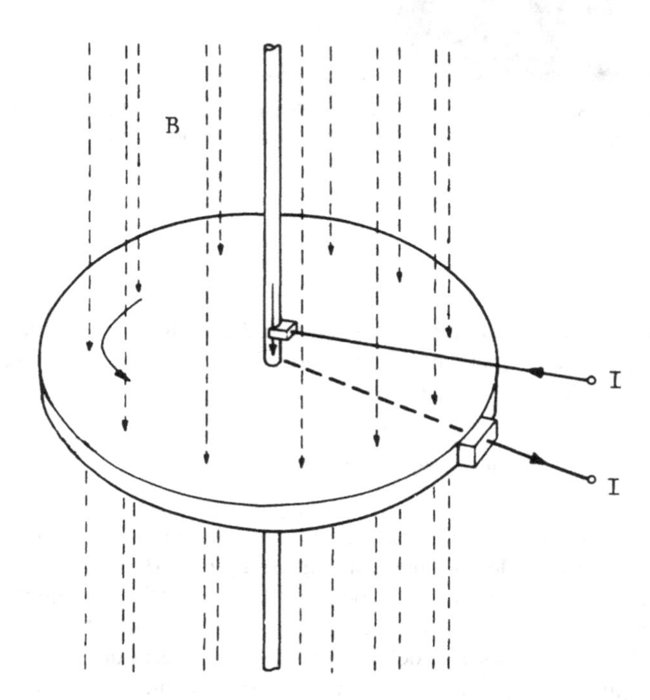

Figure 12.11. Schematic of a simple homopolar machine. The disk rotates in a magnetic field.

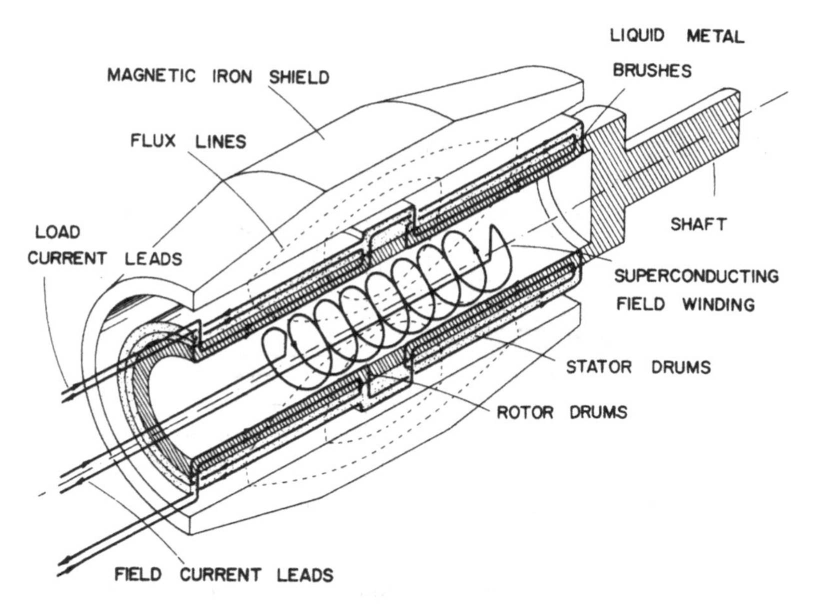

Figure 12.12. A drum type superconducting machine, developed at David Taylor Research Center.

(NMR) characteristics of the hydrogen nuclei in the body, can distinguish between the various chemical environments in which the hydrogen atoms reside in various tissues of the body. The inversion procedure used to reconstruct an image is based on the remarkable progress in tomography that has been achieved by the X-ray community. With appropriate magnetic fields and field gradients, remarkable resolution can be obtained with the MRI technique. The key element in making this technique work as well as it has is the large superconducting magnet that provides both the bias magnetic field and the magnetic field gradient. The patient is actually surrounded by the 4.2 K environment of the superconducting magnet when an MRI scan is performed on the various parts of his or her body. MRI magnets are the major commercial (nongovernmental) source of revenue for the superconducting wire manufacturers.

Magnetic Levitation

One of the most glamorous uses for superconducting magnets is in building mass transportation systems based on magnetically levitated vehicles. In fact, this is a very viable approach which the Japanese government has been developing for more than a decade. The principle on which the Japanese system works is quite simple (see Fig. 12.13). The superconducting magnets on the bottom and

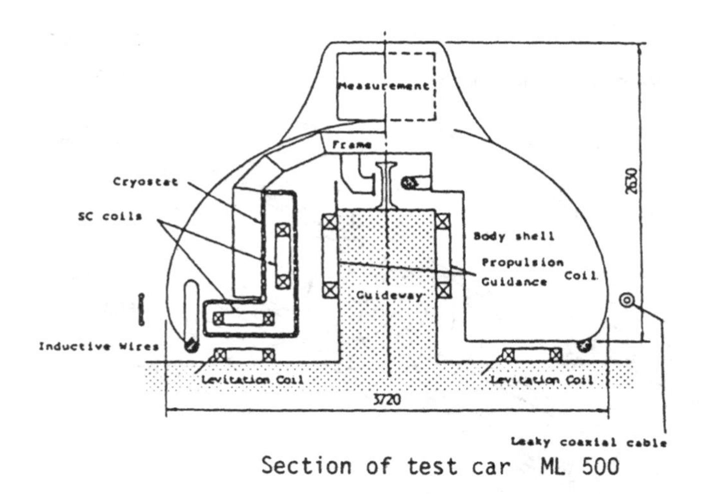

Figure 12.13. A section of the test car ML 500, developed in Japan.

side of the vehicle are located very close to the normal conducting rails. As the vehicle accelerates, eddy currents are induced in the normal conducting rails. These eddy currents produce a magnetic field that is in the opposite direction to the field produced by the superconducting magnet and provides both a levitating force from below as well as a stabilizing force from the sides so that the vehicle lifts off its wheels and is suspended both above and between the conducting rails. The propulsion of the vehicle is provided by a linear induction motor which also utilizes superconducting magnets aboard the vehicle and normal conducting magnets along the roadbed. A test version of the Japanese vehicle is shown in Fig. 12.14.

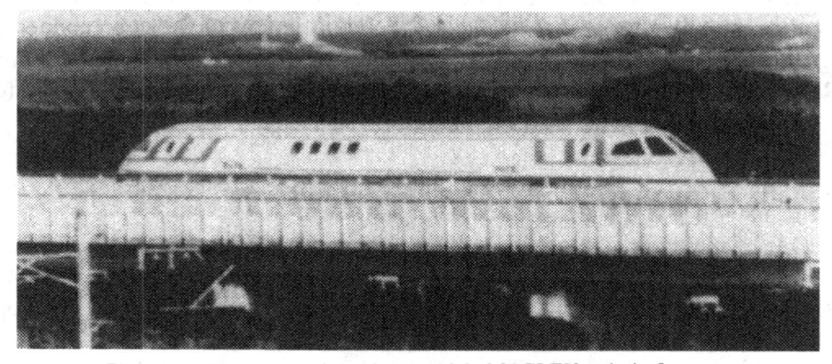

Figure 12.14. A test of the three-vehicle MAGLEV train in Japan.

Magnetic Separation

Another potential application of superconducting magnets is for the separation of magnetic materials from nonmagnetic materials or even the separation of materials by their degree of magnetism. The principle on which this method is based is that the force on a weakly magnetic particle is proportional to the product of its magnetic susceptibility and the gradient of the magnetic field. Superconducting magnets offer the possibility of high fields and high field gradients and therefore could provide very efficient separation. For example, the separation of iron ore is a process that could be made more efficient with the use of superconducting magnets.

POWER TRANSMISSION AND DISTRIBUTION

The transmission of electrical energy over large distances with very small losses is an important problem for the power industry. Superconductors are certainly ideal for this purpose since for dc currents the losses are negligible. Essentially lossless power transmission at dc is possible but of very much less interest to the power industry since it would be difficult and expensive to transform to 60 Hz at both ends. A 60-Hz superconducting cable was developed at Brookhaven National Laboratory. This effort succeeded in minimizing the losses that occurred in the stabilizing matrix because of eddy currents induced by the ac currents and solved many technological problems associated with the very large currents, the junctions to room temperature conductors, and the refrigeration. The cross section of the final conductor is shown in Fig. 12.15.

HIGH-POWER RF APPLICATIONS

Accelerator Cavities

Accelerators for high-energy physics, particularly electron accelerators, use rf cavities to accelerate the particles. The figure of merit for these cavities is the peak value of the electric field that can be obtained inside these cavities since this is directly proportional to the accelerating energy that can be imparted to the particle. This maximum rf electric field is of course, by Maxwell's equations, directly proportional to the maximum rf magnetic field that can be sustained by the cavity before it goes into the normal state. Theoretically, the peak surface magnetic field that can be obtained is the superheating field $H(\mathrm{s})$, which in most superconductors is only a little larger than the thermodynamic critical field H_c and is nowhere near H_{c2}, which determines the phase boundary for dc fields. The state of the material

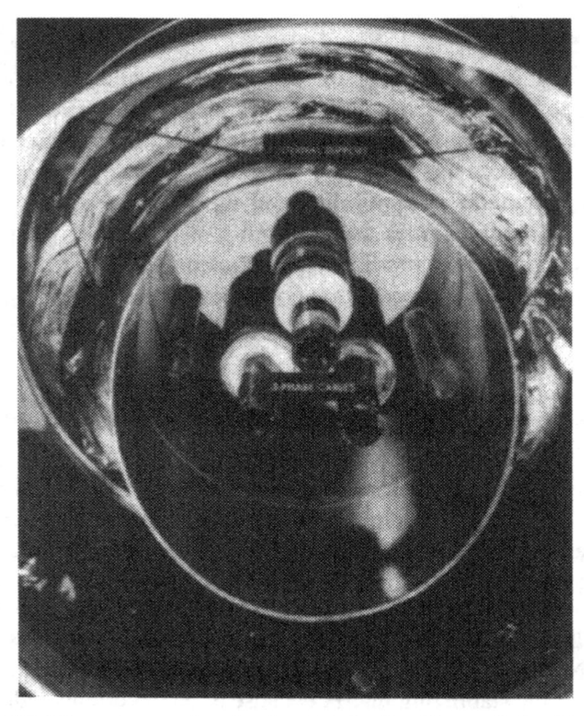

Figure 12.15. Three-phase ac line developed at the Brookhaven National Laboratory.

that has been used very successfully is Nb, and a state of the art example of a Nb accelerating structure is shown in Fig. 12.16. The reason for the odd shape of these cavities is to allow the passage of the electrons through the center, and the design maximizes the electric field along the trajectory of the electrons. The maximum electric fields that have been obtained are in the 10- to 20-MeV range, which corresponds to surface magnetic fields of several kilogauss. These values are not limited by intrinsic superconducting properties but by breakdowns that are caused by imperfections in the surface causing electron emissions which eventually short out the cavity. To even approach these field levels, the rf losses in the cavities must be minimized so that there is negligible heating of the surfaces. Again it was a difficult technological task to find ways of passivating the Nb surface to minimize the residual value of surface resistance that would maximize or enhance the quality factor of these cavities, thereby allowing the cavities to be operated at very high stored energies.

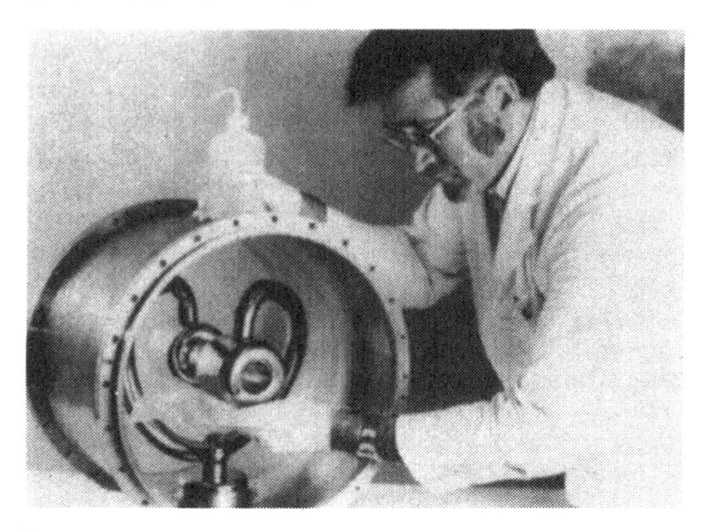

Figure 12.16. A Nb split ring resonator at Argonne National Laboratory.

RF Cavities for High-Power Filters and Multicouplers

Unlike receivers, in which the use of high-power components is not necessary, often transmitters need to have filters with very well defined bandwidths able to handle large amounts of rf power. Bulk superconducting cavities, similar to the accelerator cavities discussed above but without the complicated geometrical constraints imposed by the need to accommodate a particle beam, are an ideal approach. Filters can be designed to have flat response over some range of frequencies and very steep cutoffs on either side. This is important if there are many transmitters and receivers on the same platform (e.g., a satellite). Broad cutoffs mean excess bandwidth, which means much larger separation of channels and therefore fewer channels.

13

HIGH-T_c CUPRATES

NEW MATERIALS

A truly revolutionary series of discoveries were made from 1986 to 1988—superconductivity up to 125 K! These remarkable events were initiated by two scientists from the IBM Laboratory in Zurich, Alex Müller and Georg Bednorz. In October 1986, their paper, entitled "Possible High T_c Superconductivity in the Ba–La–Cu–O System," was published in the journal *Zeitschrift für Physik*. This paper was destined to become a milestone in the history of superconductivity, for it described the discovery of a new class of materials which displayed high-temperature superconductivity.

It may seem surprising but at first the Bednorz and Müller paper did not receive much attention. This was not because there was not a lot of interest in higher-transition-temperature superconductors but quite the contrary; there had been a large number of false alarms because of the intense interest. At first glance, the original Bednorz and Müller article could easily be taken for another blind alley. Nevertheless, their work turned out to be a real discovery.

At a meeting of the Materials Research Society (December 1986, Boston), K. Kitazawa reported important results obtained by a University of Tokyo group directed by Shoji Tanaka. The Tokyo group isolated the superconducting phase and also measured the magnetic properties of the new superconductor, confirming the Bednorz and Müller discovery.

The new material turned out to be a copper oxide, namely, a La–Ba–Cu–O compound. (See Fig. 13.2) The fact that this particular material displayed high-temperature superconductivity came as a complete surprise to scientists in the field. Previous research had been indicating that progress in increasing T_c was connected with niobium-based materials, but the breakthrough to high-temperature superconductivity came from a completely different direction. Nevertheless, the search conducted by Bednorz and Müller was quite logical. First of all,

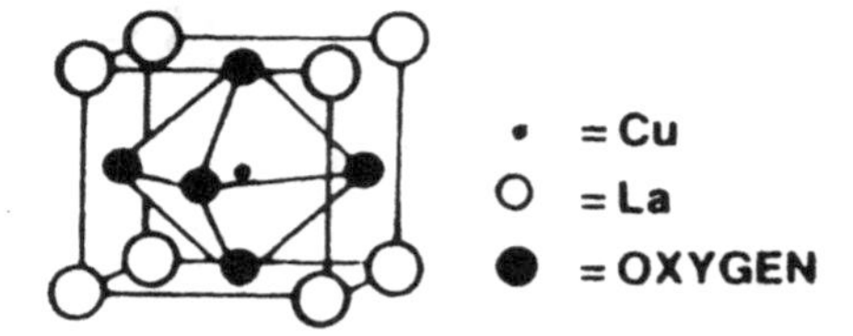

Figure 13.1. Crystal structure of the simple cubic perovskite.

they had been attracted by the unusual properties of superconducting oxides (see Chapter 9). Furthermore, they were working on a class of oxides that had been shown to be metallic by Raveau and co-workers in France and which seemed to have a crystallographic distortion driven by a strong electron–lattice interaction, called a Jahn–Teller distortion. In addition, Müller had had extensive experience working with ferroelectrics. These factors, together with phenomenal intuition, led Bednorz and Müller to their discovery.

Bednorz and Müller worked by themselves, even without a technician. This triumph of "small-scale" science, together with their innovative approach, serves as a good lesson in the history of science. The value of this lesson goes far beyond the physics of superconductivity.

In 1987, Bednorz and Müller won the Nobel Prize in physics for their discovery. Never before in the history of the Nobel Prize had there been such a short interval between the discovery and the award.

The superconducting phase was found to crystallize in the K_2NiF_4 structure, which is a layered perovskite with a strongly anisotropic crystal structure, as can be seen from Fig. 13.2. The superconductor had the composition $La_{2-x}Ba_xCuO_4$ (a 2–1–4 compound), with the superconducting properties strongly dependent on x. The optimum x from this early work seemed to be about 0.2. In rapid succession and in many laboratories, the barium was substituted by strontium and calcium, and the transition temperature was raised to nearly 40 K! The record T_c was held by the material $La_{1.85}Sr_{0.15}CuO_4$.

Several months later, groups at the Universities of Alabama and Houston under the direction of M. K. Wu and P. W. Chu, respectively, jointly announced the discovery of a 95 K superconductor consisting of a mixed phase of yttrium, barium, copper, and oxygen. It had thus become possible to observe superconductivity at the temperature of liquid nitrogen.

Their sample had a composition that was meant to exhibit the same crystallographic structure as the 40 K superconductor but, luckily, their sample contained a minority phase that had the very high T_c. Soon after the initial report, several labs (Naval Research Laboratory, IBM—Almaden, and AT&T Bell Labs) were able to isolate the phase responsible for the superconductivity and to perfect the process for preparing high-quality samples of this record-breaking material. This superconductor has the stoichiometry $Y_1Ba_2Cu_3O_7$ (abbreviated 1–2–3

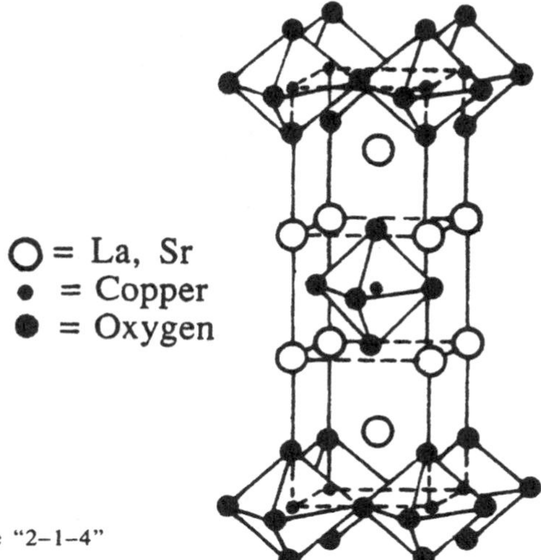

Figure 13.2. Crystal structure of the "2–1–4" $La_{2-x}Sr_xCuO_4$ compound.

compound) and the crystal structure shown in Fig. 13.3. This structure is an oxygen-defect perovskite and is very anisotropic. As can be seen from the figure, there are Cu–O planes and Cu–O chains. This structure is also based on the simple cubic perovskite (*the simple cubic perovskite structure is illustrated in fig. 13.1*) consisting of three such cells stacked one upon the other; however, the yttrium and barium are on the corner sites and alternate in a Ba–Y–Ba–Ba–Y–Ba sequence. There are strategic oxygen vacancies: no oxygens in the yttrium plane and two oxygens missing from the Cu–O layer between the barium planes. These latter missing oxygens are ordered, and the chains are very well defined so that the structure is not tetragonal, but is orthorhombic. It is slightly larger in the direction of the chains (the *b* direction) than in the perpendicular direction.

Soon after these discoveries, it was found that nearly all of the rare-earth elements, including the magnetic rare earths, could be substituted for Y without having a significant effect on the transition temperature. Two notable exceptions to this are the rare earths Ce and Pr. Nobody has been able to synthesize the 1–2–3 compound with the Ce substitution, and the Pr-substituted compound is an insulator.

In the following 12 months, many new compounds and classes of compounds were discovered. Notable among these were the Bi–Sr–Cu–O compound discovered by Raveau (CAEN, France), the Bi–Sr–Ca–Cu–O compounds discovered by Maeda *et al.* (ETL, Japan), with transition temperatures up to 115 K, and the Tl–Ba–Ca–Cu–O compounds discovered by Hermann and Sheng (University of Arkansas), and perfected by the IBM—Almaden group, with transition temperatures up to the present maximum of 125 K. These materials, some of

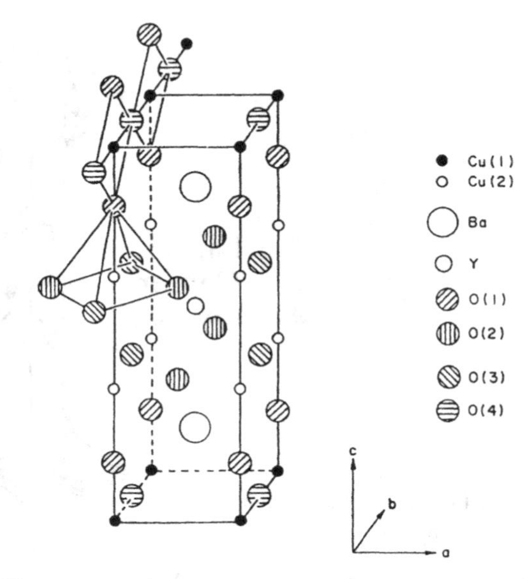

Figure 13.3. Crystal structure of the $Y_1Ba_2Cu_3O_7$ compound.

whose crystal structures are shown in Fig. 13.4, are similar to the 1–2–3 materials in that they each have one, two, or three copper oxide planes separated either by a bismuth or a thallium oxide plane. These latter planes replace the copper oxide chain that separates the two copper oxide planes in the 1–2–3 materials.

In Chapter 9, we talked about the Ba–Pb–Bi–O compound. In 1988, there appeared a paper, entitled "Superconductivity near 30K without Copper: the $Ba_{0.6}K_{0.4}BiO_3$ Perovskite" (Cava *et al.*, 1988), which reported on a new material with $T_c \approx 27$ K. Later on, T_c was raised to 34 K. This class of superconducting oxides is distinguished by a cubic structure (as opposed to a layered one). Thus, one is dealing with yet another class of superconducting oxides with a high T_c. In addition, these materials, unlike the copper oxides, do not contain magnetic ions.

There have been other new materials discovered even more recently at the University of Tokyo (Takagi *et al.*), with the most notable of these being the Nd(Ce,Pr)CuO compounds with crystal structures identical to those of the La–Ba–Cu–O superconductors discovered by Bednorz and Müller. The most important characteristic of these latest compounds is the nature of the carriers that provide the conduction. All the previous materials were hole conductors, whereas the Nd(Ce,Pr)CuO compounds are electron conductors. The critical temperature turned out to be comparatively low (~25 K). The discovery of these

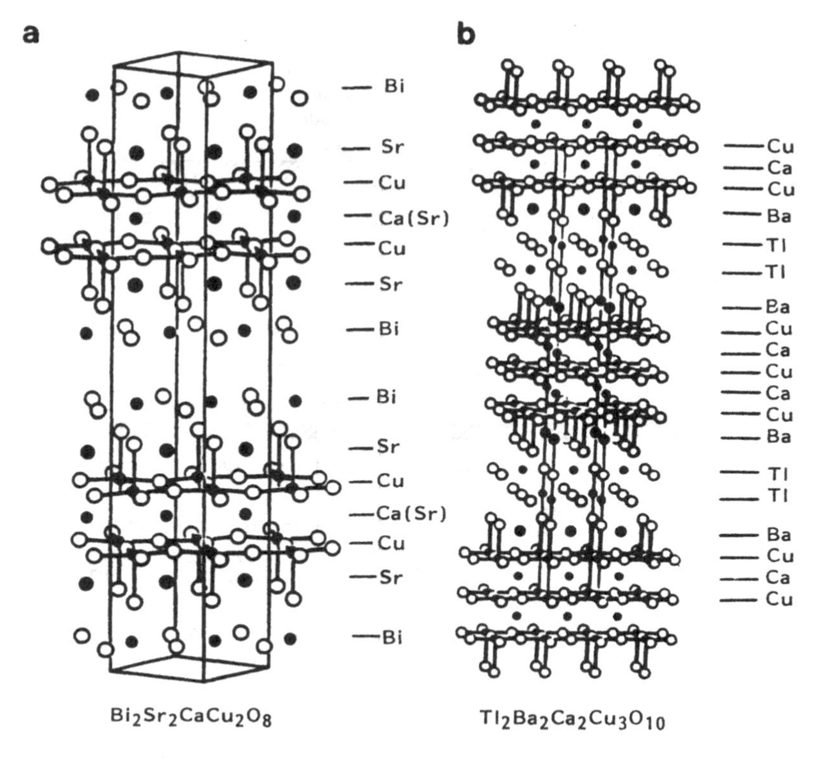

Figure 13.4. Crystal structure of (a) the $Bi_2Sr_2Ca\ Cu_2O_8$ compound and (b) the $Tl_2Ba_2Ca_2Cu_3O_{10}$ compound.

materials was important not only because they have n-type carriers, but also because they form a sort of a bridge between the cuprate family and ordinary superconductors. The temperature range of the layered cuprates is quite wide, about 100 K [from 25 K for Nd(Ce,Pr)CuO to 125 K for the Tl-based compounds]. The lower boundary of this range practically touches the upper T_c boundary of the conventional materials (for Nb_3Ge, $T_c = 22.3$ K). In addition, off-stoichiometric compositions can have T_c as low as several degrees.

The discovery of the new high-T_c oxides has resulted in an unprecedented intensity of research. Figure 13.5 shows how the critical temperature increased over the years. A similar curve would be obtained if we plotted along the vertical axis the number of papers on superconductivity. About 7000 (!) scientific papers were published during 1987–1989.

The great interest is more than just a transitory phenomenon. It is brought about by the exotic properties of these superconductors and by the great potential

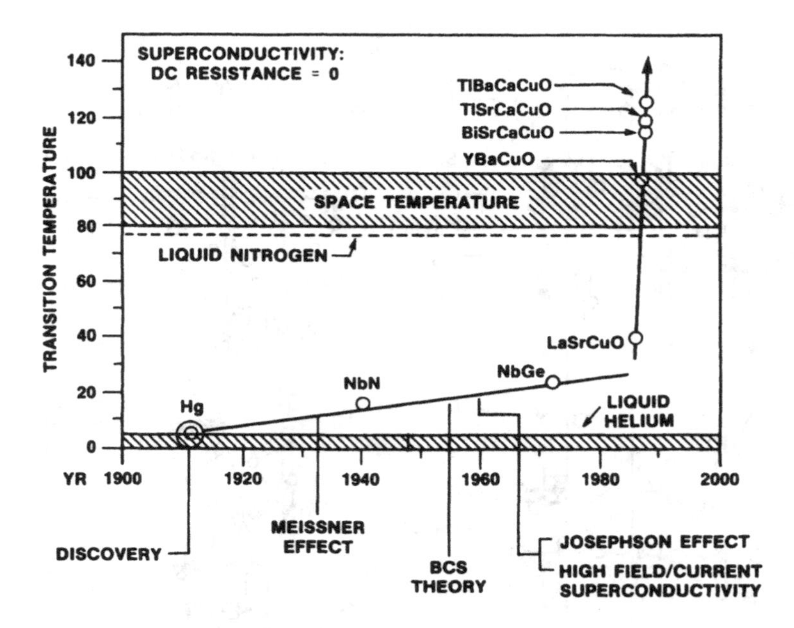

Figure 13.5. The time evolution of the superconducting transition temperature. Also plotted are significant events in the history of superconductivity.

for applications. In addition, it should be noted that Bednorz and Müller's breakthrough came in an entirely new direction. Almost nothing was known about the cuprates with regard to their normal or superconducting properties. That is why there has been and continues to be a lot of research work in this area for scientists from a wide variety of fields of specialization and having diverse research interests.

As a result of worldwide research on such a scale, there has been enormous progress in the understanding of various aspects of high-T_c superconductivity. Experiments have been focusing on the search for new materials with higher values of the superconducting parameters, such as T_c and critical current, and on the analysis of their properties. Many research centers are working on the development of applications.

When it comes to theory, it should be stressed that in the last few years, as a result of the discovery of high-T_c superconductivity, an intensive and fruitful development of the theory of superconductivity has taken place. Many different models have been proposed and developed. This process has been very beneficial to the theory of superconductivity, as well as to the entire field of solid state physics. Of course, all of the proposed models cannot be relevant to the oxides. However, even if a model is inapplicable, this does not mean that its creation was a waste of time. If this model is correct, it is still of value to the theory of superconductivity and may serve as a starting point of the search for a new superconducting system. To be appropriate for cuprates, a theory should be correct in itself and, moreover, should be relevant to the properties of the cuprates. This second factor can be judged only by experiment. The final acceptance of any theory will result from an experimental confirmation of nontrivial predictions.

There are many open questions in the physics of high-temperature superconductivity. They are concerned with the origin of high T_c, a rigorous description of the thermal, magnetic and other properties of high T_c superconductors, an analysis of their anisotropy, and so on. In this regard, we think that the La–Sr–Cu–O compound is playing an exceptionally important role. The importance is due to the relative simplicity of its structure. At the same time, the behavior of compound typifies that of the whole class of cuprates. In a way, it is playing a role similar to that of hydrogen in atomic physics.

In the next section, we are going to describe the main properties of the high-T_c oxides. Of course, our primary interest is in explaining what is unique about the new oxides, that is, what makes them different from other superconductors and other solids.

COMMON PROPERTIES OF THE CUPRATES

Normal Properties

All the copper oxides described in the previous section have some very important features in common. The crystal structures are all anisotropic, showing clear two-dimensional features; in addition, the 1–2–3 material has a well-defined one-dimensional feature, namely, the chains. This anisotropy of the crystal structure manifests itself in the electrical transport properties, which are also highly anisotropic, with the high-conductivity direction clearly parallel to the planes and due to conduction in the planes. The conduction perpendicular to the planes is less by about two to four orders of magnitude depending on the specific compound and the quality of the single crystal used to make the measurements. This demonstrates that for the most part the carriers are confined to the planes and that interlayer transitions are less favorable.

Another important feature is the fact that the Cu–O planes have to be "doped" in order to be conducting. For example, the parent material of the La–Sr–Cu–O superconductor, La_2CuO_4, is an antiferromagnetic insulator. The Cu–O planes in this case do not have any metallic characteristics. As Sr, Ba, or Ca is added to the compound, electrons are removed from the Cu–O plane, leaving behind vacancies in the band (holes). Eventually there are enough holes to make the Cu–O layers metallic, and they become superconducting, with a transition temperature that depends strongly on the carrier concentration. Similarly, the conduction in the 1–2–3 material are strongly dependent on the degree of oxidation of the chain layer. The formation of the chains removes electrons from the planes, and the maximum conduction of the planes corresponds to an oxygen stoichiometry of seven. As oxygen is removed from the layers containing the chains (see below), the conductivity of the planes decreases, the chains become partially disordered, and T_c decreases. The correct oxygen stoichiometry of the bismuth and thallium compounds also provides the optimum doping of the copper oxide planes but not in such a direct way as for the 2–1–4 and 1–2–3 materials. In the Nd(Ce,Pr)CuO materials, the Cu–O planes are doped by having an electron added to the filled band upon addition of Ce or Pr. The superconducting transition temperature for these materials is only slightly lower than for the Ba-, Sr-, or Ca-doped 2–1–4 compound.

In line with the previous comments about the doping of the planes to make these compounds both metallic and superconducting, it is important to note that the carrier concentrations of all these compounds are much smaller than for typical metals. The carrier concentrations for these materials range from 10^{21} to 10^{22} carriers per cubic centimeter. Those of typical metals are one to two orders of magnitude larger. The Hall effect measurements by which most of the carrier concentrations are derived have some unusual features as well. The Hall constant has a significant temperature dependence in the 1–2–3 compound but is much less temperature dependent for the 2–1–4 superconductor.

The thermal conductivity in these materials also differs from that in most metals. As is known, the heat flow (see Chapter 5) is made up of electron and lattice contributions, so that $\kappa = \kappa_{el} + \kappa_{lat}$. In usual metals, $\kappa_{el} >> \kappa_{lat}$. Even in the dirtiest samples, where the electron mean free path is small, and so is the corresponding heat flow, lattice thermal conductivity in the normal state never makes up more than a few percent of the total. Cuprates represent a different case [see, e.g., Jesowski *et al.* (1987)]: in them, $\kappa_{el} << \kappa_{lat}$. Heat is mainly transported by the lattice.

The electrical resistance displays a linear temperature dependence. This property has attracted considerable attention, which is not surprising, for we are dealing with an important normal-state characteristic. It should be pointed out, however, that this property is not unique to the cuprates. It is observed in some other layered conductors as well [e.g., $ZrTe_3$, 2H–TaS_2; see, e.g., the review by Bozovik (1990)]. These latter materials also display a normal resistance that is linearly dependent on temperature, without being superconductors at all or with very low T_c. In addition, this property turned out to be nonuniversal among the cuprates as well. For instance, Nd(Ce,Pr)CuO displays a quadratic dependence. The temperature dependence observed in the cuprates differs from that in normal metals because of the peculiarities of transport phenomena in materials with reduced dimensionality and low carrier concentration.

Recent photoemission experiments (Veal *et al.*, 1989; Tagahashi *et al.*,

1988) demonstrated the presence of a Fermi edge. The existence of a Fermi surface has been confirmed also by the recent observation of de Haas–van Alphen oscillations (Mueller *et al.*, 1990).

Let us mention the main difficulty with applying the common experimental tools of metal physics (Fermiology), such as cyclotron resonance, the de Haas–van Alphen effect, and magnetoresistance, to the new materials. The complications have to do with the fact that all these methods require good crystals of the normal metal at a low temperature (close to $T = 0$ K). At higher temperatures, relaxation processes become strong, and all resonance signals are washed out. However, superconductors become normal only above T_c, which is very high for the cuprates. The high critical temperatures are in this case an obstacle to the application of the traditional methods. Of course, superconductivity can be destroyed by placing the sample in a magnetic field, but the critical field of Y–Ba–Cu–O is very high (while good crystals are available for this material). Recent experiments on the de Haas–van Alphen effect employed a special explosive technique, which allowed sufficiently high magnetic fields to be achieved.

The results described here are very important, because although the materials are exotic (see Appendix E for more details), the presence of a Fermi surface shows that by definition we are dealing with a metal. After all (see, e.g., Mackintosh, 1963), metal is nothing but a solid with a Fermi surface.

We have described some properties determined by one-particle excitations (electrons, holes). In addition, many-body systems such as metals are characterized by a set of collective excitations [phonons, plasmons (see Appendix D), magnons, etc.]. An analysis of these is very important, because it is directly related to the pairing mechanism.

Neutron spectroscopy (Maraki *et al.*, 1987; Ramirez *et al.*, 1987; Boni *et al.*, 1988) has identified low-lying optical phonon modes. For example, in La–Sr–Cu–O, the modes are at $\omega_1 \simeq 100\ \text{cm}^{-1}$ and $\omega_2 \simeq 200\ \text{cm}^{-1}$. The presence of such low modes is an interesting and unique property of the cuprates. Another important feature of the lattice dynamics in these materials is the strong anharmonicity. Note that this feature is typical for ferroelectrics; this provided the motivation for Müller and Bednorz in their search for high T_c in the oxides [see also Müller (1990)]. Raman effect data (Slakey *et al.*, 1990) show that there is a continuum of electronic states in the low-energy part of the spectrum.

In addition, a layered conductor possesses a set of peculiar plasmon branches which together from a plasmon band (see Appendix D). Particularly important is the low energy part of the plasmon spectrum, which contains the phonon-like acoustic branch. This branch can provide an additional attraction between carriers and weaken the direct Coulomb repulsion.

There are also strong experimental indications of short-range magnetic fluctuations.

These data, together with information on the superconducting properties, form the basis for our theoretical understanding of the physics of high-T_c superconductivity.

Superconducting State

The cuprates display a number of properties that are in many ways similar to those of conventional BCS superconductors. First of all, flux quantization (Gough *et al.*, 1987) and Josephson tunneling experiments have demonstrated that their superconducting state is made up of paired carriers. Indeed, these experiments are directly based on the effects of carrier pairing (see Chapter 4).

Another important property of conventional superconductors is the presence of an energy gap. A number of investigations, including tunneling experiments, have shown that the cuprates also exhibit an energy gap when in the superconducting state. Photoemission spectroscopy (Imer *et al.*, 1989) revealed that the density of states has a structure similar to that which is usually seen by tunneling spectroscopy (see Chapter 3). It was found that the density of states was shifted indicating the opening of an energy gap.

There are still a lot of interesting open questions related to the behavior of the energy gap, its magnitude, its anisotropy, and the appearance of multigap structure (see below); however, the presence of the energy gap has now been established experimentally.

Just as in conventional superconductors, one observes a jump in the heat capacity at $T = T_c$ (although this jump is somewhat washed out due to fluctuations; see below), which indicates that the origin of the phase transition is similar. At temperatures close to 0 K, normal metals display linear behavior of the heat capacity (see Appendix C). This linear term is absent in the heat capacity for conventional superconductors. As far as the high-T_c oxides are concerned, there has been considerable effort directed toward trying to find an intrinsic linear term in their heat capacity near $T = 0$ K. However, by now it has been established that, just as for usual superconductors, there is no such term present (i.e., for Bi–Sr–Ca–Cu–O see R. Fischer *et al.*, 1988 and M. Sera *et al.*, 1988).

An interesting effect was observed by J. Torrance (IBM) and confirmed by other workers. He found that T_c was a nonmonotonic function of carrier concentration. There is a certain optimal value of this concentration, n_c, which corresponds to maximum T_c. When $n > n_c$, the critical temperature begins to come down, and at a sufficiently large value of n there is a superconductor-to-normal metal transition. This effect is analogous to the nonmonotonic behavior of T_c in superconducting semiconductors (see Chapter 9).

Thus far, we have listed some properties which illustrate similarities between the high-T_c oxides and conventional superconductors. On the other hand, the former also possess many exotic properties. For example, the superconducting coherence length is extremely short: conventional superconductors have coherence lengths of approximately 10^{-4} cm (see Chapter 2), whereas the cuprate superconductors have coherence lengths on the order of 10^{-7} cm. For instance, magnetic measurements indicate that in the La–Sr–Cu–O compound, $\xi_0 \simeq 20$ Å.

The short coherence length increases the significance of fluctuation effects in the vicinity of the superconducting transition. These effects have been clearly observed in heat capacity experiments, in which the sharp jump is replaced by a more gradual change with a residual tail extending above T_c [see the review by Fisher *et al.* (1988) and Ginzberg *et al.* (1988)].

Another important difference between conventional and cuprate superconductors is in the behavior of the elastic constants above and below the transition temperature. The elastic constants of conventional superconductors are unaffected by the transition whereas in the cuprates, the sound velocity, for example, is dramatically affected by the transition to the superconducting state.

Positron lifetime is another example of a physical property that is strongly affected by the superconducting transition in the cuprates, but not in conventional superconductors (Jean *et al.*, 1988). When a positron is shot into a metal, it annihilates after a finite period of time (the positron lifetime). It is well know that for an ordinary metal, transition to the superconducting state has no effect on the positron lifetime. The picture is different in the cuprates. For example, in La–Sr–Cu–O, the lifetime is longer below T_c and increases with decreasing temperature, so that the lifetime shift $\Delta\,\tau$ reaches its maximum at $T = 0$ K. The relative shift $\Delta\,\tau/\tau_n$ is as large as several percent.

A few words should be said about the electromagnetic properties of the cuprates. Because of their short coherence length, the cuprates are Type II superconductors (see Chapter 5). Therefore, they can enter a mixed state with a vortex structure. Below, we will describe the peculiar vortex dynamics in these materials.

We should also note that heat capacity measurements in the mixed state have the presence of the normal phase. In this case, $\gamma = \gamma(H)$. Clearly, $\gamma(H_{c2}) = \gamma_n$. Making a linear approximation to $\gamma(H)$ and measuring the derivative $\partial\gamma/\partial H$, one can estimate the value of γ_n. For instance, for La–Sr–Cu–O, one finds $\gamma_n = 9$ mJ/(mol K^2).

The microwave properties of the new materials are also quite interesting; they are important for practical applications. High residual losses are observed.

*We have pointed out above that the new superconductors, like conventional ones, contain paired carriers. An important question, having direct bearing on the problem of the origin of high T_c, is that of the symmetry of the pair wave function. In conventional superconductors, the pair wave function has S-wave symmetry. At present, there exists convincing experimental evidence that the same is true for the cuprates. This includes results from investigations of the temperature dependences of the penetration depth [see, e.g., Uemura *et al.* (1989)] and the Knight shift (Barrett *et al.*, 1990), which have the form given by the BCS theory, and from elegant experiments with composite [(Y–Ba–Cu–O)–Pb] rings (Little *et al.*, 1989). In the latter experiments, a persistent current was

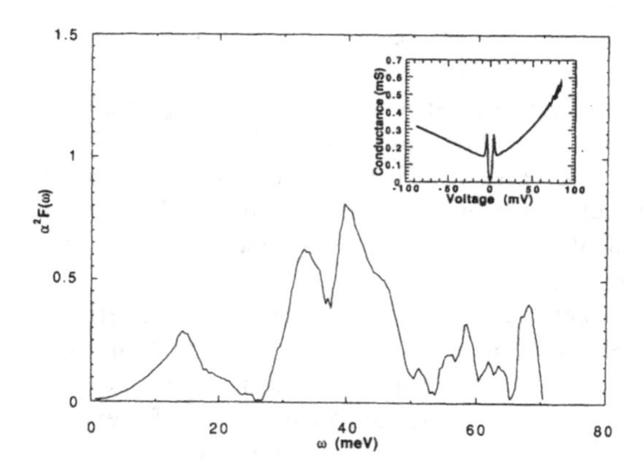

Figure 13.6. The Eliashberg function, $\alpha^2(\Omega)F(\Omega)$, for $Nd_{1.85}Ce_{0.15}CuO_{4-y}$ as obtained from point-contact tunneling spectroscopy data. The inset is the superconducting tunneling conductance which shows the sharply resolved gap structure and the low (10%) zero-bias conductance.

observed in the ring, without any suppression of the order parameter in Pb: Since Pb contains S-wave pairs, it follows that Y–Ba–Cu–O must have the same symmetry.*

Tunneling spectroscopy has proven to be the most powerful method of studying conventional superconductors. This method was described in Chapter 6. The short coherence length in the cuprates turns out to be a serious obstacle to their investigation by this technique. In this regard, the n-type materials are the most promising, because the coherence length is longer in them than in all other cuprates. The first such investigation was recently carried out at the Argonne National Laboratory by point-contact tunneling spectroscopy (Huang *et al.*, 1990). The function $\alpha^2(\omega)F(\omega)$ was reconstructed (see Fig. 13.6: $\lambda = 0.95$, $\mu = 0.1$, $T_c^{\text{calc.}} = 21$ K); it showed low-frequency phonon peaks, indicating that electron–phonon coupling is a key mechanism responsible for superconductivity in this material.

We have mentioned above that the thermal conductivity, κ, of the oxides is mainly determined by lattice heat transfer. As the temperature is lowered below T_c, κ is observed to increase. This means (see Chapter 5) that the principal relaxation mechanism is phonon scattering off electronic excitations.

The last few years have been a period of intensive developments in the theory of superconductivity. With regard to the intriguing question of the origin of superconductivity and the nature of pairing in the cuprates, it must be pointed out that the enormous experimental progress has been imposing greater and greater constraints on theories. In Appendix E, we describe an approach to the theory of high T_c developed by us (jointly with H. Morawitz).

PREPARATION OF THE CUPRATE SUPERCONDUCTORS

The remarkable progress that has been made in the last few years is, in part, due to the ease with which "ceramic" samples of many of these superconducting materials can be fabricated. The typical method for preparing polycrystalline powders and pellets of these compounds for the investigation of their properties is as follows. The appropriate proportions of the oxides or carbonates of the constituent metals are carefully ground and mixed (e.g., yttrium oxide, barium carbonate, and copper oxide). This mixture is then calcined (most of the carbonate is converted to oxide and the mixture is then mostly oxides) at a temperature above 900°C for several hours. The exact temperature depends on the particular compound. The powder is removed from the furnace and reground and reheated. This process is continued until the mixture is converted into the correct crystalline phase. The powder is then either directly annealed at low temperatures in the appropriate environment of oxygen or pressed into a pellet, rod, cylinder, or other desired form and then annealed. The most crucial step is the annealing, which must provide the right amount of oxidation to maximize the conductivity and the superconducting properties. Single crystals can be grown from the superconducting powder by melting and then very slowly cooling a mixture of the powder in an excess of CuO. The excess CuO decants from the crucible (preferably zirconia) and leaves beautiful crystals behind in the crucible. A very long oxidation protocol is required in order for these crystals to become superconducting throughout. Many crystals are "granular" superconductors because they have not been adequately oxidized. Single-crystal-like samples have also been grown from a stoichiometric melt by extremely slow cooling, and the result is a ceramic that is highly textured. This "melt-textured" superconductor has the highest critical current density of any nonthin film form of Y–Ba–Cu–O (see Francavilla *et al.*, 1990).

Thin films of these materials have been very successfully fabricated using most of the techniques described in Chapter 8 (the reader should read Chapter 8 before continuing). Some of the best-quality films have been made by an *in situ* process using either single-target high-pressure sputtering in a mixture of argon and oxygen, laser evaporation under several millitorrs of oxygen or multiple-source evaporation under a partial pressure of activated oxygen. When deposited at a substrate temperature of about 700°C, these films are quite smooth, have nearly the bulk transition temperature, and have very high ($>10^6$ A/cm^2) critical current densities measured at 77 K and zero magnetic field.

CRITICAL CURRENT DENSITIES

Ultimately, for these cuprate superconductors to be universally useful, they must be fabricated in a form that can carry large current densities at high magnetic field levels. There are electronic thin-film applications in which the magnetic field requirements are substantially relaxed. A description of these applications as well as the general prognosis for applications will be presented at the end of this chapter.

It has been extremely difficult to fabricate bulk polycrystalline samples that have technologically significant critical current densities at fields above H_{c1}. The main difficulty has been with the poor quality of grain boundaries in the bulk ceramic samples. Grain boundaries are typically where defects are refined out of the more perfect crystalline grains. In conventional superconducting materials such as NbTi and Nb_3Sn, these grain boundaries tend to pin the flux quanta; this has the effect of improving the high-field current-carrying capabilities of these materials. For the cuprate materials, the opposite is true; because of the very short coherence lengths, the grain boundaries act like very weak Josephson junctions, strongly attenuating the maximum supercurrent that can be transported across the boundary. The Josephson critical current is very strongly affected by small magnetic fields so that the critical current densities of this type of material are extremely low in a modest magnetic field (1 T). The aim of current research is to reduce the number of grain boundaries in the direction of the current flow by preparing samples with very long grains that are aligned with the *c*-axis perpendicular to the current flow. These highly textured samples have the best high-field properties for bulk samples (see above).

Thin-film samples have been shown to have properties at zero field that are better than those of conventional superconductors. The current densities for thin-film samples in moderate fields are now considerably better than those of the bulk materials and probably reflect the fact that thin films can be grown with highly oriented, extremely clean grain boundaries. The properties of even these films degrade at high temperatures, and this is almost certainly due to the weak pinning potentials that are naturally available to pin the flux lattice (the maximum pinning energy per vortex line per unit length is proportional to the coherence length squared) because of the extremely short coherence lengths that are intrinsic to these materials. Table 13.1 summarizes some of the most recent critical current densities for bulk and thin-film samples.

Table 13.1. Best Critical Currents of Cuprate Superconductors (Transport Measurements)[a]

Sample	Magnetic field along *c* (T)	Critical current (A/cm²)	
		$Y_1Ba_2Cu_3O_7$	$Bi_2Sr_2Ca_1Cu_2O_{10}$
Film at 77 K	0	$\sim 10^7$	5×10^6
	10	10^5	$<10^2$
Bulk at 77 K	0	5×10^4	5×10^3
	10	3×10^3	<10

[a]Francavilla, 1990 and McConnell and Wolf, 1989.

CRITICAL MAGNETIC FIELDS

The upper critical field H_{c2} in the BCS theory can be written as

$$H_{c2}(0) = \pi^2\Delta^2(0)/2e^2\hbar v_F^{\,2}$$

where $\Delta(0)$ is the value of the superconducting energy gap at zero temperature. From this expression, it can be seen that the upper critical fields in these materials might be extremely high because of the small Fermi velocity and the large energy gap. In fact, critical field measurements have been hampered not only by the fact that the measurements can be done only near T_c, where the fields are accessible in the laboratory, but also by the lack of significant flux pinning in the same temperature range, making the traditional resistance determination totally unreliable. Magnetization measurements near T_c have only recently given reliable numbers for this very important quantity. Furthermore, the critical fields are quite anisotropic, being much larger when the field is applied parallel to the planes than when it is applied perpendicular to the planes. This is again a consequence of the anisotropy of the structure and, in particular, the anisotropy of the transport. Since most of the conduction is in the planes, a magnetic field applied parallel to the planes will not be very effective in destroying superconductivity within the planes. Reliable measurements for the upper critical field $H_{c2}(0)$ only for Y–Ba–Cu–O: H_{c2} with the field perpendicular to the A–b plane is about 130 T; with the field parallel to the a–b plane, H_{c2} is about 650 T (see Welp *et al.*, 1989). $H_{c2}(0)$ for La–Sr–Cu–O is known more approximately: it is about 80 T perpendicular to the Cu–O planes and 240 T parallel to the planes.

STATUS OF APPLICATIONS

There are several areas in which the cuprates may have a technological impact in the next few years.

Thin films have reached a level of performance, in terms of both their critical current density and their surface resistance, that makes them suitable for many of the passive microwave components described in the section on superconducting electronics in Chapter 12. Components such as filters, delay lines, transmission lines, and switches have been made and are performing much better than any nonsuperconducting counterpart. These components are significantly better than their copper, gold, and silver counterparts operated at the same cryogenic temperature. In fact, for such devices, the cuprate superconductors should ultimately outperform even conventional superconductors because of the larger energy gap in the cuprates, which allows fewer quasiparticle excitations.

The remaining technical problems are ones of scale-up and reproducibility in a manufacturing environment.

SQUID development is another technological area in which the cuprate superconductors may have a significant impact in the very near future. (SQUIDs are described in the electronics applications section in Chapter 12.) Recently, SQUIDs using "weak-link" type junctions formed at the grain boundaries in Tl–Ba–Ca–Cu–O films have exhibited very low noise performance on several devices. This is a proof-in-principle that we have a technological breakthrough in this area.

To date, there have been no reports of high-quality Josephson or single-particle tunnel junction devices operating at 77 K. However, several types of junctions of the superconductor–normal metal–superconductor (SNS) type have been fabricated from Y–Ba–Cu–O with silver, gold, or damaged superconductor as the normal metal, and these junctions have worked very well at low temperature ($T<60$ K), but none have worked at 77 K (see McConnell and Wolf, 1989).

Vortex flow devices have recently been proposed as an alternative to Josephson junction devices for active elements in microwave circuits. These devices rely on vortex propagation through a very narrow constriction in a superconducting film to provide a variable inductance that can provide a phase shift in a coupled transmission line or can provide gain when driven by an input element that injects vortices in the constriction. This vortex flow transistor (Martens *et al.*, 1989) has been made using Tl–Ba–Ca–Cu–O films that have very low vortex pinning and are therefore ideal for this device.

Simple applications involving bulk materials such as magnetic shields and current leads are presently feasible but are not significant enough to warrant a significant investment in technology. In fact, these applications will probably be dependent on the development of a useful conductor that can operate at temperatures higher than can be achieved with conventional superconductors. When this hurdle is overcome, the large-scale power applications described above will become feasible.

APPENDIXES

BOUND STATES IN TWO DIMENSIONS

Consider a two-dimensional potential well $V(\rho)$. Writing the Schrödinger equation in polar coordinates (ρ, φ) and assuming that there is an energy level with $E \ll U$ (this assumption will be justified by the solution), we obtain

$$\frac{1}{\rho} \frac{\partial}{\partial \rho} \left(\rho \frac{\partial \psi}{\partial \rho} \right) = \frac{2m}{\hbar^2} V \tag{A.1}$$

Integrating over $d\rho$ from 0 to ρ_1 [where $d \ll \rho_1 \ll \varkappa^{-1}$; d is the radius of the potential well, and $\varkappa = (2m|E|)^{1/2}/\hbar$] and equating the logarithmic derivatives inside and outside the well [see Landau and Lifshitz (1977) for details of the soltuion], we find

$$|E| \sim \left(\frac{\hbar^2}{ma^2} \right) e^{-\hbar^2/m\lambda^*} \tag{A.2}$$

where $\lambda^* = \int_0^\infty V\rho d\rho$. Note the similarity between Eq. (A.2) and the expression for the energy gap in Eq. (2.2)

THE METHOD OF ELEMENTARY EXCITATIONS (QUASIPARTICLES)

The quasiparticle method is one of the basic techniques of modern quantum theory. It is indispensable for studying the low-lying excited states of systems of many interacting particles. We illustrate this method by discussing phonons, which are quasiparticles describing lattice vibrations.

A crystal lattice consisting of ions undergoing small-amplitude vibrations can be viewed as a set of independent harmonic oscillators. The energy of a harmonic oscillator of frequency Ω is equal to $E = \hbar\Omega(n + ½)$. The energy of lattice vibrations is then given by

$$E = \sum_{q_i} \hbar\Omega_{q_i} \left(n_{q_i} + \frac{1}{2} \right) \tag{B.1}$$

The term $0.5\ \Sigma_{q_i}\hbar\Omega_{q_i}$ is the energy of zero-point vibrations. Its origin is in the quantum-mechanical uncertainty principle, which states that even in the lowest-energy state the ions cannot rest in their equilibrium positions. The zero-point energy cannot be given up by the lattice and does not have to be considered when analyzing thermal motion.

Equation (B.1) tells us that the lattice thermal energy cannot be an arbitrary number but has a discrete set of values determined by the integers n_{q_i}.

When calculating thermodynamic and other properties of the crystal, it turns out to be very convenient to visualize the oscillating system as made up of a set of so-called elementary excitations, or quasiparticles, whose energies are determined by Eq. (B.1). Instead of speaking about an oscillator of frequency Ω_{q_i} in energy level n_{q_i}, we can consider n_{q_i} quasiparticles, each of energy $h\Omega_{q_i}$. They are called phonons, or "sound quanta." The phonon picture reflects the quantum character of lattice vibrations. This concept allows us, instead of having to analyze a complicated system of vibrating ions, to deal with an intuitive and simple model of a gas of quasiparticles with definite energies and momenta.

The interaction of electrons with thermal lattice vibrations determines many nonequilibrium properties of metals, such as electrical and thermal conductivities. Analysis of electron scattering by thermal vibrations reduces, in the quasiparticle method, to the problem of collisions between particles of two kinds. Electron–phonon collisions are subject to the usual conservation laws.

Historically, the problem of lattice vibrations was the first to involve analysis of a many-particle system. Nowadays, quasiparticles are introduced in many other cases as well. In its general form, the quasiparticle method was formulated by L. D. Landau in his work on superfluidity of liquid helium.

Quasiparticles describe weakly excited states of systems with many degrees of freedom, such as crystals, plasmas, electromagnetic fields, liquid helium, ferromagnetics, superconductors, and so on. The excited state of a system differs from the ground state in its energy, momentum, and possibly some other quantities (for example, when a ferromagnetic is excited, its magnetic moment changes). Quasiparticles serve to describe this excited state. One associates them with the system as a whole and assigns to them the proper values of energy, momentum, etc. While the number of quasiparticles is small, which is the case at low temperatures when the degree of excitation is low, they can be treated as an ideal gas. Thus, a quasiparticle is not a particle in the usual sense of the word but is a convenient representation of an excited state. In this representation, the problem of interaction between systems with many degrees of freedom reduces to that of interacting quasiparticles. This makes it very convenient to utilize the energy and momentum conservation laws. For example, photons describe excitations of the electromagnetic field, which has a discrete set of energy levels. Similarly, phonons describe a sound wave. Many other quasiparticles have been introduced, such as plasmons, magnons, excitons, and polarons.

ELECTRONS IN METALS. FERMIOLOGY

Quantum theory of the solid state was born in 1928, when Sommerfeld applied Fermi–Dirac statistics to electrons in metals. His model became known as the "free-electron model" and led to a number of important results.

In reality, electrons strongly interact with the lattice ions. Many processes (electrical and thermal conduction, superconductivity, etc.) depend strongly on this interaction. It is also impossible, within the free-electron model, to understand why solids are divided into metals, insulators, and semiconductors. The answer to this fundamental question is given by quantum-mechanical analysis of electrons moving in a crystal lattice.

The electron–lattice interaction leads to the formation of band structure. An energy band is an interval of allowed values of the electron energy. The electron energy spectrum in a solid is given by a function $E(\mathbf{p})$, where $\mathbf{p}$ is the crystal momentum (quasimomentum). This function (often called the dispersion relation) is, generally, quite complicated and strongly depends on the crystal structure.

If a band contains a small number of electrons, the energy near an extremum may be expanded in powers of the crystal momentum; in a cubic crystal, $E = p^2/2m^*$, where m^* is the so-called effective electron mass. Thus, the energy of a crystal electron may, under certain circumstances, be cast into a free-electron-like form. The only difference is that the real electron mass has been replaced by an effective mass, m^*. Thus, the effect of the periodic field of the lattice is to alter the electron mass.

Consider now the case when the band is almost completely filled. In this case, it is more convenient to look at the vacant, rather than filled, states. Let us suppose that every unfilled state is occupied by an electron together with a positively charged particle of the same mass and charge magnitude. Instead of a band containing a small number of vacant levels, we are now dealing with an equivalent system made up of a completely filled band and a small number of positive charges, known as holes. The latter occupy levels which were vacant in the original picture. If we now apply a weak external electric field, it will generate current, made up of electron and hole components. However, it is easy to see that the electronic component in our case vanishes, since a completely filled band does not contribute to conduction. (The field is weak enough so that the energy it imparts to an electron is less than the distance to the higher-lying unfilled band.) We thus arrive at the conclusion that a nearly filled band is equivalent to a set of holes, that is, particles whose behavior is described by the same formulas as that of electrons in an almost empty band, but with a positive charge.

It goes without saying that our introduction of the concept of holes does not mean that they really exist. This is nothing but a convenient way of describing an electron system whose behavior may be so radically altered under the influence of the crystal lattice as to become identical to the motion of a set of positively charged particles (holes).

The most efficient way to analyze the anisotropy of a metal and to evaluate its normal-state parameters is to describe the system in momentum space; this approach is called "Fermiology." Fermiology is a well-known method in the physics of metallic systems. The function $E(\mathbf{p})$ represents the dispersion relation for an electron (or a hole), and the equation $E(\mathbf{p}) = E_F$ defines the Fermi surface. For a three-dimensional isotropic system with a simple quadratic dispersion relation $E(\mathbf{p}) = p^2/2m^*$, the Fermi surface is a sphere. Anisotropy of a system is reflected in the shape of its Fermi surface.

Fermiology encompasses several different approaches. One is based on band structure calculations: the Fermi surface parameters are calculated. Another approach looks at certain experimental data which are sensitive to the topology of the Fermi surface; special theoretical analysis then allows its shape to be reconstructed. Ultrasound attenuation in a magnetic field is an example of such an experiment: attenuation oscillates as a function of the field, with the period of oscillation directly related to the geometry of the Fermi surface and its parameters. The parameters of the Fermi surface of many metals and compounds have been established with this kind of approach, and detailed results have been obtained.

This latter approach to Fermiology is similar in concept to that of tunneling spectroscopy (see Chapter 6), where theoretical analysis allows one to use tunneling data in order to evaluate the major parameters of the electron–phonon interaction.

The case of a layered system is quite important, because the new high-T_c oxides, such as La–Sr–Cu–O, have a layered structure. The dispersion relation $E(\mathbf{k}, p_z)$ is highly anisotropic (here $\mathbf{k}$ is the two-dimensional in-plane crystal momentum, and the z axis is perpendicular to the layers). In this case, the Fermi surface is cylindrically shaped. We will consider the Fermiology of a layered conductor in more detail in Appendix E.

In discussing Fermiology, one ought to mention the so-called Fermi-liquid effects. The fact of the matter is that the strong Coulomb correlation makes it necessary to consider metallic electrons not as a quantum gas, but rather as a quantum liquid. The theory of Fermi liquids was developed by Landau in 1956 on the basis of the theory of collective excitations (see Appendix B). According to the Landau theory, the low-lying excitations of a Fermi liquid can be classified in the same way as those of a Fermi gas. However, there exist some high-frequency phenomena which require that Fermi-liquid effects be taken into account, namely, the fact that quasiparticle energy is a functional of the distribution function.

Finally, let us consider the heat capacity of an electron gas at low temperatures. As the electron gas is heated, some electrons near the top of the distribution make transitions to states above E_F. The number of these electrons is approximately given by NkT/E_F (N is total number of electrons). The increase in energy of one of them is equal to the average thermal energy, that is, $\Delta E \approx kT$. Thus, the energy of the entire electron system increases by an amount proportional to the square of the temperature:

$$E \simeq \frac{(k_B T)^2}{E_F} N$$

The low-temperature heat capacity of the electron gas, $c_V = (\partial E/\partial T)_V$, turns out to be directly proportional to the temperature:

$$C_V \simeq \left(\frac{Nk_B^2}{E_F} \right) T \tag{C.1}$$

This relation is an important result of the quantum theory of metals. For $T \to 0$, the electronic heat capacity decreases linearly.

However, when one takes into account the interaction of the electrons with thermal vibrations of the lattice, this leads to deviations from the linear law [Eliashberg, 1963; Kresin and Zaitsev, 1976; see also the book by Grimvall (1981)]. The degree of this deviation is related to the strength of the electron–phonon coupling.

PLASMONS

Plasmons are not a new concept in solid-state physics. In a certain sense, they are similar to phonons. Both describe collective vibrational motion, but phonons represent vibrational motion of the lattice, whereas plasmons correspond to collective vibrational motion of the carriers relative to the lattice. This motion represents charge fluctuations and is due to Coulomb correlations in the metal (Goldman, 1947; Bohm and Pines, 1951).

The energy and the dispersion relation of plasma excitations depend strongly upon the dimensionality of the system. In an ordinary three-dimensional metal, the plasma frequency is usually high ($\hbar\omega_p \simeq 5-10$ eV) and is described by the dispersion law $\omega_p = \omega_0 + aq^2$, where $\mathbf{q}$ is the three-dimensional momentum. There is a peculiar case (discussed in Chapter 7) when there are two overlapping energy bands ("light" and "heavy" carriers); this situation gives rise to an acoustic branch ("demons").

The picture is different for a two-dimensional electron gas (Stern, 1967). In this case, there is no gap at $q = 0$, and the dependence $\omega_p(q)$ has the following form: $\omega \sim q^{1/2}$.

Consider now the case of layered conductors. This case is important, because the new high-T_c oxides belong to this class. Layered conductors are different from both an isotropic three-dimensional crystal and a two-dimensional electron gas. Here we are dealing with a highly anisotropic crystal; interlayer screening interaction is important even if interlayer carrier hopping is neglected. One can show that plasmons in layered conductors form a so-called plasmon band (see Fig. D.1). The dependence $\omega_p(\mathbf{q})$ has the following form (Kresin and Morawitz, 1990):

$$\omega_p = v_F q_\parallel \left[1 + \frac{\alpha^2}{4(\alpha + 1)} \right]^{1/2} \tag{D.1}$$

where

$$\alpha = \frac{4me^2}{\epsilon_M q_\parallel} \frac{\sinh(q_\parallel d_C)}{\cosh(q_\parallel d_C) - \cos q_Z d_C} \tag{D.2}$$

Here $0 < q_z < \pi/d_c$, where d_c is the interlayer distance, $q_\parallel$ is the in-plane momentum and ϵ_M is the background dielectric constant. For $\alpha \gg 1$ and $\epsilon_M = 1$, one obtains (Fetter, 1974): $\omega_p = v_F q_\parallel (1 + \alpha/4)^{1/2}$. Therefore, plasmon energy levels fill an entire region limited by two boundaries: the upper corresponds to

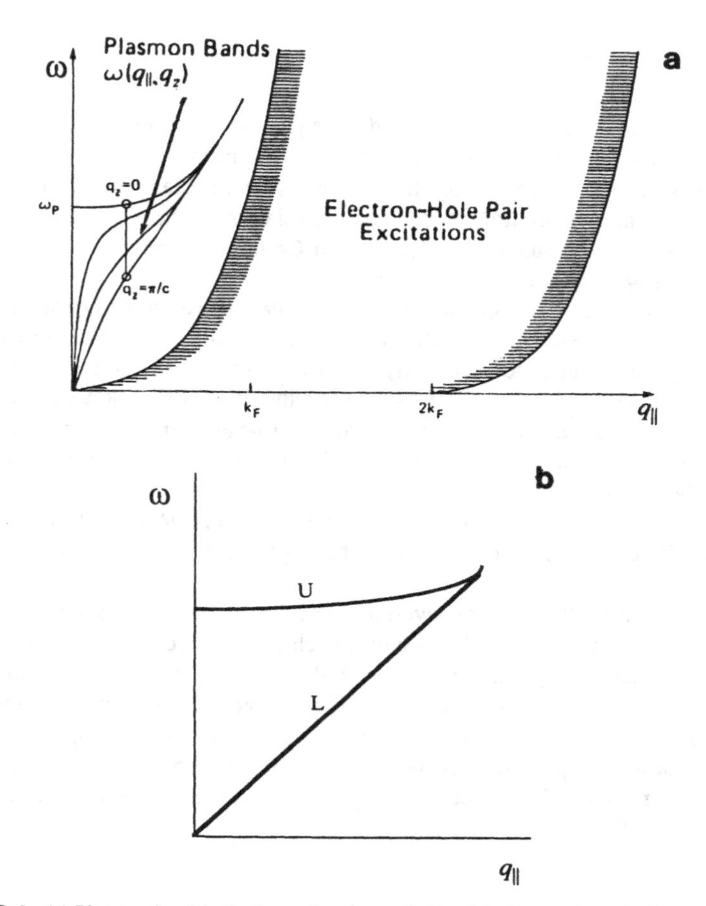

Figure D.1. (a) Plasmon band in the layered compound. Density of states is peaked near the upper and lower boundaries. (b) Qualitative picture of the plasmon spectrum in the layered conductor: U is the upper branch and L is the lower acoustic branch.

$q_z = 0$, and the lower to $q_z = \pi/d_c$. One can prove that the density of states is peaked near the boundaries; as a result, in a simplified qualitative picture, the plasmon band in a layered crystal can be visualized as a set of two branches: an upper branch (U) which is similar to the usual three-dimensional plasmons and a peculiar low-lying (L) branch with an acoustic dispersion relation (see Fig. D.1).

The presence of the L plasmon branch should also affect electron (positron) energy loss spectroscopy of the cuprates. This kind of spectroscopy is a well-known technique for detecting plasmons in ordinary metals. When a charged particle passes through the crystal, it loses its energy by exciting plasmons; experimentally, one observes resonances at the corresponding energies. However, in the usual case the plasmon energies are very large (5–10 eV), so that $\hbar\omega_p \gg k_B T$. As a result, the energy loss picture does not depend on temperature. The picture in the cuprates is different. In the absence of a gap at $q = 0$ in the plasmon dispersion relation, the inequality $\hbar\omega_p \gg k_B T$ is no longer valid. This implies that the contribution of the L branch to inelastic electron scattering should be temperature dependent (the total losses $\gamma_L \sim T^3$).

WHAT IS UNIQUE ABOUT THE HIGH-T_c OXIDES?

(Major Parameters and the Mechanism of High-T_c Superconductivity)

Since the discovery of high-temperature superconductivity, we have been actively involved, jointly with H. Morawitz (IBM Almaden Research Center, California), in formulating a theoretical framework in which to understand the normal and superconducting properties of the cuprates. We have developed a unified approach allowing the correlation of the normal and superconducting properties of the new materials. We are continuing this work and hope to describe a more refined version of the theory in the next edition of this book. In this appendix, we are going to describe our concept in the theory of high T_c [for a more detailed description, see our papers (1987–1990)].

There are several related, but nevertheless distinct, problems in the theory of high T_c: (1) analysis of the normal-state properties, (2) analysis of those properties of the superconducting state which are independent of the mechanism of superconductivity, and (3) elucidation of the origin of high T_c. In what follows, we will discuss these three aspects, paying particular attention to their interrelations.

We begin by discussing the normal-state properties of the new materials and, following that, address their superconductivity.

NORMAL-STATE PROPERTIES

One should try to understand the normal state before going on to an analysis of the superconducting properties, for a number of reasons. First of all, one is dealing with a phase transition, so one needs to understand the states between which the transition takes place. Moreover, many superconducting parameters are expressed in terms of normal-state ones, so one has to know the latter. Finally, it is essential to know the character of both single-particle and collective excitations in the normal state. The type and properties of these excitations have a direct bearing on the nature of the superconducting state.

There is an interesting contrast between the development of the physics of the cuprates and that of the physics of conventional superconductors. Just before the creation of the BCS theory, the normal-state properties of conventional metals were very well understood, but superconductivity was not. The situation with the cuprates was just the opposite: at the time high-T_c superconductivity was discovered, there already existed a good understanding of the phenomenon of superconductivity, but the normal-state properties of the cuprates were practically unknown. Therefore, before we can judge the efficacy of any theory to explain superconductivity in the cuprates, we must develop an understanding of their normal-state properties.

The cuprates become metallic, and then superconducting, as a result of carrier doping. In this regard, one can distinguish two different directions in the physics of high T_c. One is the problem of carrier doping and the dynamics of the transition from the insulating to the metallic state, and the other addresses a description of the metallic state. We are going to focus on the latter. It is important that this metallic phase undergoes a transition into the superconducting state; as a result, our analysis is directly related to the origin of high T_c.

The most important question connected with the normal state has to do with the spectroscopy of this state. In other words, we would like to understand the different excitations in the layered metallic oxides and their spectra, that is, their dispersion relations. It is known that a many-body quantum system is characterized by single-particle and collective excitations. Electrons and holes are examples of single-particle excitations, whereas such excitations as phonons, plasmons, excitons, and magnons are collective.

One-Particle Excitations

Our goal is to evaluate the major parameters, such as the Fermi energy E_F, the effective mass m^*, the Fermi velocity v_F, and the Fermi momentum K_F, which all describe the properties of one-particle excitations.

The results of our analysis are given in Table E.1 for La–Sr–Cu–O (the method will be described below). For comparison, Table E.1 also includes typical parameters of conventional metals. The low values that we find for E_F and v_F seem to us to be exceptionally important properties of the high-T_c oxides.

Let us give a brief description of our approach. The high-T_c cuprates are characterized by large anisotropy caused by the presence of low-dimensional

Table E.1.

Quantity	Conventional metals	$La_{1.8}Sr_{0.2}CuO_4$
m^*	1–15 m_e	$5m_e$
k_F (cm^{-1})	10^8	3.5×10^7
v_F(cm/s)	$(1–2) \times 10^8$	8×10^6
E_F(eV)	5–10	0.1

substructures, such as layers or chains. The most efficient way to analyze the anisotropy is to describe the system in momentum ("reciprocal") space; our method is based on Fermiology (see Appendix C). The anisotropy is reflected in the fact that the topology of the Fermi surface, $E(\mathbf{p}) = E_F$, is nonspherical.

Here we are going to focus mainly on the La–Sr–Cu–O compound (the parameters of Y–Ba–Cu–O will be described later, when we talk about the two-gap model; see also Chapters 3 and 13). We think that this oxide is exceptionally important for an understanding of the physics of high T_c and can be treated as a test case. Its importance is due to the relative simplicity of its structure. At the same time, its behavior typifies that of the whole class of cuprates. In addition, there exist highly reliable experimental data (see below) for this material.

La–Sr–Cu–O has a layered structure, with the interlayer distance much greater than the in-plane lattice period: $d_c >> d_a, d_b$. The dispersion relation $E(\mathbf{p})$ is highly anisotropic. (Here the quasimomentum $\mathbf{p} = (\mathbf{k}, p_z)$, where $\mathbf{k}$ is the two-dimensional momentum in the layer plane, and the z-axis is perpendicular to the layers.) As a first approximation, one can assume the Fermi surface to be cylindrically shaped, which corresponds to neglecting interlayer transitions. Of course, the latter will lead to small deviations from the cylindrical shape. It is important to stress that we are not assuming the Fermi curve (defined as the cross section of the Fermi surface with the plane p_z = const.) to be a circle. It turns out that one can estimate the values of the Fermi energy E_F and the effective mass m^* (defined below) without specifying the exact shape of the Fermi curve. In the case of a cylindrically shaped Fermi surface, the dispersion relation does not depend on p_z: $E = E(\mathbf{k})$. This approximation is justified by the large anisotropy of normal conductivity. Note that this approach is applicable to hole–carrier as well as electron–carrier types of materials. Hole surfaces located in the corners of the first Brillouin zone can be, by a simple translation in reciprocal space, put together into a cylinder.

Let us estimate, in this approach, the values of normal and superconducting parameters.

*The total energy can be written as

$$W = 2 \int d\underset{\sim}{k}\, dp_z E f (2\pi\hbar)^{-3} \tag{E.1}$$

Here f is the Fermi function; integration over p_z is restricted by $|p_{z\,max}| = \pi/d_c$. With the use of Eq. (E.1), we obtain the following expression for the Sommerfeld constant $\gamma = C_{el}/T$: $\gamma = (\pi/3\hbar_2)m^* k_B^2\, d_c^{-1}$, where the average effective mass is defined as

$$m^* = (2\pi)^{-1} \int d\,l\, v_{\perp}^{-1} \tag{E.2}$$

and the integration is taken over the Fermi curve: $E\kappa) = E_F$, p_z = const.; $v_{\perp} = (\delta E/\delta\kappa)_F$. In the case of a simple parabolic band, this mass is equal to the usual effective mass. The expression for γ can be used to determine m^*. The expression for E_F can be obtained in an analogous manner.

Note that for an isotropic system (spherical Fermi surface) the density of states v_F, and hence the Sommerfeld constant, is proportional to $m^* p_F$ (where p_F is the Fermi momentum, which depends on the carrier concentration), whereas for a layered structure (cylindrical Fermi surface), $v_F \sim m^* d_c^{-1}$. This enables us to obtain a one-to-one correspondence between the Sommerfeld constant and the effective mass [see Eq. (E.3)].*

Our analysis is based on the following formulas for the effective mass m^* and the Fermi energy:

$$m^* = (3\hbar^2/\pi)\, k_B^{-2}\, d_c\, \gamma \tag{E.3}$$

$$E_F = (\pi^2\, k_B^2/3)\, n/\gamma \tag{E.4}$$

Equations (E.3) and (E.4) express m^* and E_F in terms of the experimentally measured quantities d_c, n, and γ. It is important to stress that these formulas have been derived specifically for the case of a layered conductor and are different from their counterparts for an isotropic three-dimensional metal.

Consider now the Fermi momentum K_F and the Fermi velocity v_F. These quantities depend strongly on the shape of the Fermi curve and on the direction. We can estimate the values of K_F and v_F by making the isotropic approximation for the Fermi curve $E(\kappa) = E_F$;that is, by taking $K_F = (2m^* E_F)^{1/2}$ and $v_F = K_F/m^*$.

*The values of γ and n can be determined from heat capacity data [see, e.g., Phillips *et al.* (1987) and the review by Fisher *et al.* (1988)] and Hall effect measurements [see, e.g., Ong *et al.* (1987)]. It is important to stress that since the Fermi surface describes the ground state of the system, Eqs. (E.3) and (E.4) correspond to γ in the low-temperature region, near T = 0 K [that is, γ in Eqs.

(E.3) and (E.4) is in fact γ (0)]. This remark is essential because γ depends strongly on T (see below). Note also that Hall effect measurements of La–Sr–Cu–O do not display strong temperature dependence. This is due to the relatively simple band structure of this system, which allows us to employ the carrier concentration obtained by Hall measurements in La–Sr–Cu–O. The situation in the case of Y–Ba–Cu–O is different because of its more complicated band structure.

Determination of $\gamma(0)$ is not a simple task, because the system is in the superconducting state and its critical field is large. This problem has been solved (Phillips *et al.*, 1987) by analyzing the dependence of γ on H, so that $\gamma(0) \simeq (d\gamma/dH)H_{c2}$. There is, naturally, some uncertainty in the experimental data, which propagates into the numbers quoted below (Table E.1). The important issue, however, is the order of magnitude of the physical quantities. For example, the smallness of E_F and v_F that we find basically overwhelms the experimental uncertainties. Note also that the value H_{c2} = 80–100 T for La–Sr–Cu–O an be evaluated in a self-consistent way (see below).*

As a result, we can calculate the major normal-state parameters of the high-T_c materials; the results are presented in Table E.1.

One can see that the new materials contain a set of "heavy" carriers. The Fermi momentum is smaller but does not differ drastically from that in conventional metals. This fact is important, because it provides large phase space for pairing. Overall, it can be said that the values of m^* and p_F are not radically different from those found in usual metals.

What is really remarkable is the values of the Fermi energy and the Fermi velocity. They turned out to be almost two orders of magnitude smaller than in conventional metals. Instead of a Fermi sea, we really have a Fermi puddle!

We think that the small values of E_F and v_F, along with the large anisotropy, are key features of the cuprates. These small values are important for the unified approach described here.

The small values of E_F and v_F can explain a number of exotic normal and superconducting properties (the latter will be discussed later). For example, the electronic thermal conductivity in the normal state is proportional to $E_F K_F$ [see, e.g., the book by Geilikman and Kresin (1974)]. A small value of E_F should lead to a significant decrease in the electronic contribution to the total thermal transport. As was mentioned in Chapter 13, this is observed experimentally: contrary to the conventional picture, thermal conductivity in the cuprates is dominated by the lattice, so that $\kappa_{lat} >> \kappa_{el}$.

Collective Excitations

Having discussed the properties of single-particle excitations, we now focus on collective excitations, specifically, on phonons and plasmons.

We mentioned in Chapter 13 that neutron data have shown that the cuprates contain low-lying anharmonic optical modes. We believe that these modes are important for the superconducting pairing; we shall come back to this question below.

Let us now consider plasmons, another kind of collective motion. A layered conductor, such as the copper oxide we are dealing with, possesses a peculiar plasmon spectrum. We discussed its character in Appendix C. There are plasmons with low energies and an acoustic dispersion law (see Fig. D.1). It is important that the slope of this acoustic dependence is directly related to the Fermi velocity v_F, so that $\omega = s\kappa$, $s \simeq V_F$. As we have stressed above, the Fermi velocity in the cuprates is small (see Table E.1). As a result, they contain the additional acoustic L branch, which could be called "electronic sound," similar to the usual phonon branch (the speed of the electronic sound is not much higher than that of usual sound). However, this plasmon branch reaches larger energies; this fact is important to the physics of superconductivity.

This plasmon branch represents a dynamic part of the direct carrier-carrier interaction and provides "anti-screening," which weakens the Coulomb repulsion.

Both phonons and the "electronic sound" contribute to the pairing in the cuprates.

SUPERCONDUCTING STATE

We turn now to the superconducting state of the new materials. First of all, we shall consider properties that are not directly related to the mechanism of pairing; we wish to stress that they are to a large extent determined by the exotic normal-state parameters described above (see Table E.1).

Superconducting Parameters

According to our analysis, the Fermi velocity v_F is small relative to its value in conventional metals. For La–Sr–Cu–O (see Table E.1), $v_F \simeq 8 \times 10^6$ cm/s. Such a small value of v_F together with a high value of T_c results in a short coherence length. Indeed, if we use the expression $\xi_0 = 0.18 h v_F / k_B T_c$ together with the derived value of v_F, we obtain $\xi_0 \simeq 25$ Å.

*One can use a more precise definition: $\xi_0 = \hbar v_F / \pi\Delta(0)$. As is known, the energy gap $\Delta(0)$ is directly related to T_c, so that $\Delta(0) = a k_B T_c$; the value of a depends on the strength of the coupling (in the weak-coupling BCS approximation, $a = 1.76$).

Making use of Eq. (6.3) and the characteristic phonon frequency, 15 meV, we find that for La–Sr–Cu–O, $\Delta(0) \simeq 2.5 k_B T_c$. Many tunneling experiments find that indeed in La–Sr–Cu–O, $\alpha \simeq 2.5$. The uncertainty in some tunneling data probably is due to energy gap anisotropy. Using this value of $\Delta(0)$ and the values of v_F (see above) and T_c, we find the coherence length to be $\xi_0 \simeq 20$ Å.*

This short coherence length is due partly to the large energy gap, but mainly to the small Fermi velocity. The shortness of the coherence length is a very important feature of the new materials.

Using the calculated value of the coherence length and the expression $H_{c2} = (\Phi_0 / 2\pi\xi^2)$, one can evaluate H_{c2} for La–Sr–Cu–0. Here ξ is the Ginzburg–Landau coherence length: $\xi_{G-L} = a\xi_0\,[1-(T/T_c)]^{-1}$. In the weak-coupling approximation, $a = 0.74$. Strong-coupling effects increase a, so that for La–Sr–Cu–O, it becomes equal to 0.95. Using this value of a and $\xi_0 \simeq 20$ Å, we obtain $H_{c2}(0) \simeq 90$ T. It is interesting to note that a value of $H_{c2} = 88$ T was used above in order to determine the value of the Sommerfeld constant $\gamma(0)$. The fact that the calculated coherence length leads to a nearly identical value of H_{c2} illustrates the self-consistency of our approach.

Pairing in the High-T_c Oxides

The small value of the Fermi energy also has a strong impact on the superconducting properties. Let us look, first of all, at the ratio $\Delta(0)/E_F$, which is an important superconducting parameter. As a result of a superconducting transition and electron pairing, the Fermi surface undergoes a reconstruction within a layer of thickness Δ. Therefore, the parameter $\Delta(0)/E_F$ shows what fraction of the electronic states are directly involved in pairing. In conventional superconductors, this ratio is small ($\sim 10^{-4}$), whereas in the high-T_c oxides E_F and Δ are comparable: $\Delta(0)/E_F \simeq 10^{-1}$. The small value of this ratio in ordinary superconductors means that only a small number of states near the Fermi surface are involved in pairing. The picture is different in the oxides. The large value of the ratio corresponds to a significant fraction of the carriers being paired up. Naturally, this implies a short coherence length.

The possibility of having a large value of Δ/E_F and a short coherence length is directly related to the quasi-two-dimensional structure of the cuprates. Indeed, in conventional superconductors ($\Delta/E_F << 1$), pairing can occur only near the Fermi surface (Cooper's theorem). The states on the Fermi surface form a two-dimensional system in momentum space. This is important, because in two dimensions any attraction will lead to the formation of a bound state (see Appendix A). The layered structure of the cuprates makes pairing possible even for states that are distant from the Fermi surface.

The quantity $\Delta(0)/E_F$ also determines the extent of the region around T_c in which fluctuations of the order parameter manifest themselves. Fluctuational coupling for $T > T_c$ is revealed first of all in the behavior of the heat capacity. The large value of $\Delta(0)/E_F$ leads to unusual critical behavior near T_c. This was predicted by Deutscher (1987) on the basis of the small value of E_F, and then observed experimentally (see Chapter 13).

Positron Annihilation Lifetime

We mentioned in Chapter 13 that, unlike the superconducting transition in conventional superconductors, the superconducting transition in the cuprates is accompanied by an observable shift in the positron annihilation lifetime. It turns out that this effect is also due to the small value of E_F and the consequent large value of the parameter $\Delta/(0)E_F$.

Positrons entering a metal undergo a two-stage process of thermalization followed by annihilation. The metal contains bound and delocalized groups of electrons; in the high-T_c oxides, positrons annihilate mainly with the localized group, which is more populated. Clearly, the state of these electrons is not affected by the superconducting transition. However, the delocalized carriers screen the positron–bound electron interaction. In this way, the state of the delocalized subsystem also affects the annihilation process. Superconducting pairing affects the screening and therefore may affect the annihilation dynamics. In other words, since the annihilation lifetime τ is determined by the overlap of the positron and bound electron wave functions, one can expect that the superconducting transition will affect the lifetime: the shape of the positron wave function in a many-body system such as a metal may be noticeably affected by the delocalized sybsystem of the carriers.

The observed shift in τ is caused by pairing; it appears that $\Delta\tau/\tau \approx \gamma\,(\Delta/E_F)^2 \ln\,(E_F/\Delta)$ $(\gamma \simeq 1)$. The small value of E_F along with the large value of the energy gap leads to a sizable effect in the high-T_c oxides. Unlike the carriers in conventional superconductors, a significant portion of the carriers in the oxides are paired, which results in a noticeable change in screening.

Multigap Structure in the Cuprates

The short coherence length makes it possible to observe multigap structure in the high-T_c oxides (see Chapter 3).

Multigap structure is caused by the presence of overlapping energy bands; each band is characterized by its own energy gap. Differences in densities of states, pairing interactions, and other properties lead to distinct values of the gaps. Overlapping energy bands are common in conventional metals including superconductors. Nevertheless, the properties of conventional superconductors have been described accurately by a one-gap model. This is due to their long coherence length ξ_o; that is, the inequality $l << \xi_o$ (l is the mean free path), which holds for most conventional superconductors, leads to averaging, because of interband scattering. As a result, a one-gap model provides an adequate description.

The cuprates are characterized by short coherence lengths, and, as a result, it is perfectly realistic to observe multigap structure in the high-T_c oxides. Of course, this requires that different energy bands overlap. Such a situation exists in Y–Ba–Cu–O and probably in Bi–S–Ca–Cu–O. The La–Sr–Cu–O compound is characterized by a relatively simple band structure and can be described by a one-gap model. The weak temperature dependence of the Hall coefficient is a manifestation of such a structure.

The situation for Y–Ba–Cu–O is entirely different; it contains two different energy bands, of a quasi-two-dimensional and a quasi-one-dimensional nature. The applicability of the two-band picture is manifested in the strong temperature dependence of the Hall effect and is supported by band structure calculations (Pickett, 1989). The Fermi surface in this system consists of a cylindrical piece and two planar pieces.

As a result, one should observe two energy gaps in the Y–Ba–Cu–O superconductor. Several experiments have in fact indicated the presence of two energy gaps. Tunnelling *I–V* characteristics (Gurvitch *et al.*, 1989) clearly show two distinct gap values. The most convincing evidence comes from NMR experiments. NMR data (Warren *et al.*, 1987) were interpreted in the framework of the two-gap model. A more recent detailed investigation (Barrett *et al.*, 1990) of the temperature dependence of the Knight shift (see Chapter 5) clearly demonstrated the presence of two energy gaps. Therefore, Y–Ba–Cu–O contains two superconducting subsystems: two distinct bands with different energy gaps.

As was described in Chapter 3, a two-band superconductor is characterized, generally speaking, by three independent coupling constants: λ_a, λ_b, and λ_{ab}. For Y–Ba–Cu–O it is natural to assume that $\lambda_b = 0$. Note that although $\lambda_{ab} > \lambda_{ba}$ $(\lambda_{ab}/\lambda_{ba} = \nu_b/\nu_a)$, the effective contribution of λ_{ab} is small; it is diminished by a large renormalization factor, so that it enters as $\lambda_{ab}(1 + \lambda_a + \lambda_{ab})^{-1}$. The negligibly small value of λ_b is due to strong screening of electron–phonon scattering.

If we take $2\Delta_1(0)/T_c = 6$ and $2\Delta_2(0)/T_c = 3.5$ (these values are consistent with the tunneling and NMR data), it follows that $\tau = \Delta_2/\Delta_1 = 0.6$. Let us assume by analogy with La–Sr–Cu–O that $\lambda_a = 2.5$; note that the ratio $T_c/\tilde{\Omega}$ has approximately the same value for both oxides ($\tilde{\Omega}$ is the characteristic phonon frequency: $\tilde{\Omega} = 15$ meV for La–Sr–Cu–O, and $\tilde{\Omega} = 40$ meV for Y–Ba–Cu–O, so that $T_c/\tilde{\Omega} = 0.25$). One can prove that $\tau = \lambda_{ab}(1 + \lambda_a)\lambda_a^{-1}$; as a result, we obtain $\lambda_{21} = 0.4$. Therefore, the superconducting state in the chains is caused by weak interband coupling, whereas the planes are characterized by strong coupling.

*Based on the two-band model, one can estimate the parameters of Y–Ba–Cu–O.

According to heat capacity data, the Sommerfeld constant for Y–Ba–Cu–O is equal to $\gamma = 16$ mJ/(mol·K^2). In the presence of two bands, $\gamma = \gamma_{pl} + \gamma_{ch}$. The Fermi surface of Y–Ba–Cu–O consists of a cylindrical part and a set of planes. The cylindrical part, corresponding to the quasi-two-dimensional (planes), is similar to the quasi-two-dimensional band of La–Sr–Cu–O. Using the relation in Eq. (A.9) and assuming that the value of the effective mass m^* of the carriers in the planes is the same as for La–Sr–Cu–O ($m^* = 5\ m_e$), we obtain $\gamma_{pl} = 0.04$ mJ/(cm^3·K^2). Correspondingly, $\gamma_{ch} = 0.15$ mJ/(cm^3·K^2). Assuming also that the hole concentration is similar to that in the $La_{1.85}Sr_{0.15}CuO_4$ compound ($n_h = 3 \times 10^{21}$ cm^{-3}), we obtain $\kappa_{F;\ pl} = (2\pi n d_c)^{1/2} = 4.7 \times 10^7$cm^{-1}, and $v_{F;\ pl} = 10^7$ cm/s. The Fermi energy turns out to be equal to $E_F = 0.2$ eV; a similar value has been obtained by Deutscher and us (1988) by analyzing the critical region. It is remarkable that photoemission data (Arko *et al.*, 1989) also give the same value for the Fermi energy.

Using the values of the parameters obtained above, we can estimate the in-plane coherence length $\xi_{pl} = \hbar v_F / \pi \Delta(0)$, and we obtain $\xi_{pl} = 10$ Å. Let us consider the carriers in the quasi-one-dimensional band (chains). Using the value of γ_{ch} obtained above, we can evaluate the Fermi velocity, obtaining $v_{F;\ ch} = 5 \times 10^6$ cm/s. This leads to a very small value of the coherence length: $\xi_{ch} = 7$ Å. The carrier concentration, n_{ch}, is unknown. In the presence of two bands, the Hall effect depends strongly on temperature and cannot be used to determine n directly. It is probable that $m^*_{ch} >> m^*_{pl}$; for example, if we took $n_{ch} = n_{pl} = 3 \times 10^{21}$, we would obtain $m^*_{ch} = 25 m_e$.*

ORIGIN OF HIGH T_c (PHONON–PLASMON, OR GENERALIZED PHONON, MECHANISM)

As was noted above, many experimental data indicate that the new high-T_c oxides display many features of the BCS theory, such as carrier pairing and the presence of an energy gap. However, we are dealing with an "exotic" version of the theory, brought about by a combination of such factors as the layered structure and the small values of the Fermi energy and velocity.

The pairing is caused by some intermediate field. In the conventional BCS theory, this field is formed by phonons. But generally speaking, as we discussed in Chapter 6, attraction can be mediated by other excitations, such as plasmons, excitons, and magnons. One has to figure out which of these excitations are responsible for interelectron attraction.

The question of how high T_c is created by virtual exchange of some excitations may be broken up into two distinct parts: (1) the question of the existence of a particular excitation, and (2) the question of whether the coupling of the carriers to this excitation is sufficient to provide the high T_c that is observed. For phonons, the answer to (1) is obvious. No one doubts their existence. Recent experiments (see above) have demonstrated that the answer to this question for plasmons is also positive. Therefore, when speaking of phonon and plasmon exchange, one has to realize that these excitations are not hypothetical, but rather very real.

It should be pointed out that plasmons in the cuprates are quite peculiar. Their spectrum is similar to that of phonons, and they look like additional branches of the phonon spectrum. Together, phonons and the acoustic quasi-two-dimensional plasmons comprise the low-frequency part of the spectrum.

Let us look first at the role of phonons. There is now massive evidence pointing at the important role played by the electron–phonon interaction. This includes, for instance, the behavior of the thermal conductivity for $T < T_c$ which we mentioned earlier (see Chapter 13), which proves that the carrier–lattice interaction is the main relaxation mechanism, and tunneling experiments (Gurvitch *et al.*, 1989), which indicate that the location of a distinctive feature on the current–voltage curve coincides with a peak in the phonon density of states. In principle, we are dealing with a system which contains low optical phonon modes with an anharmonic lattice (see above, Chapter 13), so that it is natural to expect a strong electron–phonon coupling.

One sometimes comes across the argument that the small value of the isotope effect coefficient, found in oxygen isotope experiments, reveals that the electron–phonon interaction is weak in these compounds. The fallacy of this argument follows from the complicated nature of the isotope effect (see Chapter 6). It is certain that the cuprates exhibit the isotope effect. Interestingly, it turns out to be even stronger for carrier concentrations lower than n_{max} [where $T_c(n_{max}) = T_{c\ max}$]. It is clear, however, that in the case of such complicated multicomponent lattice structures one may not directly correlate the magnitude of the shift and the coupling strength.

It is thus crucial to be able to determine the intensity of the coupling between the carriers and the lattice. This problem is intimately related to that of establishing the strength of the binding force between carriers in a Cooper pair. Indeed, the critical temperature is determined by the strength of the coupling, λ, and by the energy scale of the excitations, W: $T_c = T_c(\lambda, W)$ (see Chapter 6). As a result, the question of the strength of the coupling is very important. For example, if coupling is weak, then the observed high T_c can be provided by a large value of the energy scale W; the latter should be much greater than the phonon energies, and the phonon contribution will not be important. On the other hand, if the coupling is strong, then phonons must play an essential role. Tunneling spectroscopy is the most powerful method of determining the coupling constant; we described this method in Chapter 6. However, a number of factors, in particular the shortness of the coherence length, make tunneling spectroscopy of the new materials very difficult.

Nevertheless, one can use a different method to estimate the magnitude of the coupling constant. This method is based on analysis of heat capacity data.

Electron–phonon interaction leads to a deviation of the electronic heat capacity $C_{el} = \gamma T$ from a linear law; in other words, the Sommerfeld constant depends on temperature. Near T = O K, $\gamma(0) = \gamma^b(1 + \lambda)$, where γ^b is the so-called band value, corresponding to the frozen lattice.

In other words, in the low-temperature region the carriers are "dressed" by phonons. Thermal motion leads eventually to the "undressed" band value. The coupling constant λ is equal to $\lambda = [\gamma(0)/\gamma^b] - 1$.

Hence, if $\gamma(0)$ and γ^b are known, the coupling constant can be calculated. The value of $\gamma(0)$ for La–Sr–Cu–O has been determined by a unique method utilizing low-temperature measurements of the heat capacity in a magnetic field (see above). The value of γ^b can be determined from the electronic heat capacity in the high-temperature region. The high critical temperatures of the cuprates allow one to determine γ^b from the jump in the heat capacity [one should use Eq. (6.4), based on the theory of strong coupling]. As a result, one finds that in La–Sr–Cu–O, $\gamma_{ph} \simeq 2.25$–2.5.

We thus come to the important conclusion that the electron–phonon interaction in La–Sr–Cu–O is strong. The value of $\gamma_{ph} \simeq 2.5$ that we have obtained for La–Sr–Cu–O is large, meaning that the electron–phonon interaction plays an important role. Note for comparison that, for example, in Pb, which is considered a strong-coupling superconductor, $\gamma_{ph} \simeq 1.4$ (see Chapter 6).

The electronic structure of Y–Ba–Cu–O is more complicated than that of the La–Sr–Cu–O compound. This is related to the presence of the additional quasi-one-dimensional band (see above). As a result, there are three, rather than one, coupling constants. The values of these constants are given on page 217. Let us remark that the higher T_c of Y–Ba–Cu–O is due to two factors. Mainly, it is important that here the optical phonon frequencies are higher (with the same value of the coupling contant λ_{aa}); besides, the very fact that there are two different bands present is favorable for coupling.

Hence, the cuprates are characterized by strong carrier–lattice coupling. It is natural to ask the question, is this coupling sufficient to explain the observed high T_c? It turns out that the coupling constant is still not large enough. Indeed, we should ask ourselves, how large should γ_{ph} be in order to provide the measured T_c of the cuprates? This question can be answered with the use of Eq. (6.8).

Using $T_c \simeq 40$ K and $\tilde{\omega} \simeq 15$ meV (this value was obtained from neutron spectroscopy data), we find $\lambda \simeq 5$ (for $\mu^* \simeq 0.1$; larger values of μ^* result in even greater λ values). Note that if we take $\tilde{\omega} \simeq 10$ meV, this will also result in a larger coupling constant. Hence, λ should be greater than about 5 in order to provide for the observed high value of T_c. Therefore, there is need for an additional mechanism.

This conclusion raises the question, what *is* this additional mechanism? We think that the additional attraction is mediated by the peculiar low-lying acoustic plasmon branch (see above). An "electronic" phonon represents a natural addition to the phonon spectrum. As a result, we can speak of a generalized phonon mechanism.

Note that the presence of this additional plasmon mechanism does not imply that λ_{pl} is large. The contribution of a mechanism to T_c depends on the coupling strength λ and on the energy scale W. The coupling constant λ_{pl} of the pairing interaction of the carriers with plasmons may be small, with the sizable enhancement of T_c being due to the larger scale of plasmon energies. This can be seen directly from the equation

$$T_c \simeq T_c^{ph} \, (\tilde{\omega}_{pl}/T_c^{ph})^{\nu} \tag{E.5}$$

Here $\nu = \lambda_{pl}(\lambda_{ph} + \lambda_{pl})^{-1}$, T_c^{ph} is given by Eq. (6.8), and $\tilde{\omega}_{pl} = \langle\omega\rangle_{pl}$.

Equation (E.5) gives T_c in the presence of both modes: phonons and two-dimensional acoustic plasmons. The observed critical temperature $T_c \simeq 40$ K corresponds to $\lambda_{ph} \simeq 2.25$–2.5, $\lambda_{pl} \simeq 0.25$, and $\tilde{\omega}_{pl} \simeq 60$ meV.

Thus, high T_c in the cuprates is caused by the interaction of carriers with low-lying collective excitations of lattice (phonons) and electronic ("electronic" sound) origins (i.e., by a phonon–plasmon, or generalized phonon, mechanism).

SUGGESTED READINGS

MONOGRAPHS

A. Abrikosov, 1988, *Fundamentals of the Theory of Metals* North-Holland, Amsterdam.

A. Abrikosov, L. Gor'kov, and I. Dzyaloshinski, 1963, *Methods of Quantum Field Theory in Statistical Physics* Dover, New York.

A. Barone and G. Paterno, 1982, *Physics and Applications of the Josephson Effect*, Wiley, New York.

N. Bogoliubov, N. Tolmachev, and D. Shirkov, 1959, *A New Method in the Theory of Superconductivity*, Cons. Bureau, New York.

A. Campbell and J. Evetts, 1972, *Critical Currents in Superconductors*, Taylor, London.

A. Cracknell and K. Wong, 1973, *The Fermi Surface*, Clarendon, Oxford.

P. de Gennes, 1966, *Superconductivity of Metals and Alloys*, Benjamin, New York.

B. Geilikman and V. Z. Kresin, 1974, *Kinetic and Non-Steady Effects in Superconductors*, Wiley, New York.

V. Ginzburg and D. Kirzhnits, 1982, *High Temperature Superconductivity*, Cons. Bureau, New York.

G. Grimvall, 1981, *The Electron-Phonon Interaction in Metals*, North-Holland, Amsterdam.

I. Khalatnikov, 1982, *An Introduction to the Theory of Superfluidity*, Benjamin, New York.

I. Kulik and I. Yanson, 1972, *The Josephson Effect in Superconductive Tunneling Structures*, Israel Program for Scientific Translations, Jerusalem.

E. Lifshitz and L. Pitaevskii, 1980, *Statistical Physics*, Oxford, New York Vol. II.

F. London and M. London, 1950, *Superfluids*, New York, Vol. 1.

A. Maradudin, E. Montroll, and G. Weiss, 1963, *Theory of Lattice Dynamics in the Harmonic Approximation* (Supplement to Solid State Physics), Academic, New York.

A. Migdal, 1967, *Theory of Finite Fermi System and Application to Atomic Nuclei*, Interscience, City.

D. Pines, 1963, *Elementary Excitations in Solids*, Benjamin, New York.

D. Saint-James, G. Sarma, and E. Thomas, 1969, *Type II Superconductivity*, Pergamon, Oxford.

R. Schrieffer, 1900, *Theory of Superconductivity*, Benjamin, New York.

D. Shoenberg, 1952, *Superconductivity*, Cambridge University Press, London.

M. Tinkham, 1975, *Introduction to Superconductivity*, McGraw-Hill, New York.

T. Van Duzer and C. Turner, 1981, *Principles of Superconductive Devices and Circuits*, Elsevier, New York.

E. L. Wolf, 1985, *Principles of Electron Tunneling Spectroscopy*, Oxford University Press, Oxford.

REVIEWS

A. Abrikosov and I. Khalatnikov, 1959, Theory of Superconductivity, *Adv. Physics* **8,** 45.

M. Beasley and T. Geballe, 1984, Superconducting materials, *Physics Today* **37,** 60.

D. Brewer (ed.), 1966, *Quantum Fluids,* North-Holland, Amsterdam.

F. Dyson, 1970, Future of physics, *Physics Today* **23,** 204.

B. Geilikman, 1966, The electron mechanism of superconductivity, *Sov. Phys.—Usp.* **16,** 141; 1973, Problems of high-temperature superconductivity in three-dimensional systems, *Sov. Phys.—Usp.* **16,** 17.

B. Geilikman, 1975, Adiabatic perturbation theory for metals and the problem of lattice stability, *Sov. Phys.—Usp.* **18,** 190.

A. Goldman and S. Wolf (eds.), 1984, *Percolation, Localization and Superconductivity* (Nato ASI), Plenum, New York.

C. Gorter (ed.), 1964, *Progress in Low Temperature Physics,* North-Holland, Amsterdam, Vols. 1–4.

R. Greene and P. Chaikin, 1984, Organic superconductors, *Physica B* **126,** 431.

D. Gubser, T. Francavilla, S. Wolf, and J. Leibowitz (eds.), 1980, *Inhomogeneous Superconductors* (AIP Conference Proceedings No. 58) AIP, New York.

H. Gutfreund and W. A. Little, 1979, The Prospects of Excitonic *Superconductivity in Highly Conducting One-Dimensional Solids,* J. Derreese, R. Evrard, and V. van Doren (eds.), Plenum, New York.

J. Hulm, C. Jones, R. Muller, and T. Tien, 1967, Low Carrier Concentration Superconductors, Proceedings of the Tenth International Conference on Low Temperature Physics, Moscow p. 86.

A. Mota, P. Visani, and A. Pollini, 1989, Magnetic Properties of Proximity Induced Superconducting Copper and Silver, *J. Low Temp. Phys.* **76,** 465.

R. Parks (ed.), 1969, *Superconductivity,* Marcel Dekker, New York.

G. Stewart, 1984, Heavy fermion systems, *Rev. Mod. Phys.* **56,** 755.

H. Weber (ed.), 1977, *Anisotropy Effect in Superconductors,* Plenum, New York.

J. Wheatley, 1975, Experimental Properties of Superfluid ^{3}He, *Rev. Mod. Phys.* **47,** 415.

S. Wolf and V. Kresin (eds.), 1987, *Novel Superconductivity,* Plenum, New York.

N. Zavaritskii, 1977, Electron-phonon interaction and characteristics of metal electrons, *Sov. Phys.—Usp.* **15,** 608.

REFERENCES

Because of the nature of this book, we have not provided a comprehensive list of references. More comprehensive listings of references for Chapters 1–12 can be found in the SUGGESTED READINGS and for Chapter 13 in the *Journal of Superconductivity* Vol. 2, No. 1 and Vol. 3, No. 1. The references below are usually examples for each topic. In addition, the articles we referred were either not reflected in the review articles or monographs or were milestones in the development of the field.

P. Allen and R. Dynes, 1975, *Phys. Rev. B* **12,** 995.

P. Anderson, 1959, *J. Phys. Chem. Solids* **11,** 26.

J. Bardeen, L. Cooper, and J. Schrieffer, 1957, *Phys. Rev.* **108,** 1175.

N. Bogoliubov, 1947, *J. Phys. USSR* **11,** 23; *Sov. Phys.—JETP*, **7,** 41.

K. Chang, M. Dacorogna, M. L. Cohen, J. Mignot, G. Chouteau, and G. Martinez, 1985, *Phys. Rev. Lett.* **54,** 2375.

J. Clarke, 1969, *Proc. R. Soc. London, Ser. A* **308,** 447.

D. Deaver and W. Fairbanks, 1961, *Phys. Rev. Lett.* **7,** 143.

G. Eliashberg, 1961, Sov. Phys.—JETP **13,** 1000; 1963, **16,** 780.

A. Fetter, 1974, *Ann. Phys. (N.Y.)* **88,** 1.

R. Fisher, S. Kim, B. Woodfield, N. Phillips, L. Taillefer, K. Hasselbach, J. Flouquet, A. Giorgi, and J. Smith, 1989, *Phys. Rev. Lett.* **69,** 1411.

T. L. Francavilla, V. Selvamanickam, K. Salama, D. H. Liebenberg, *Cryogenics* (to be published).

H. Frohlich, 1950, *Phys. Rev.* **79,** 845.

H. Frohlich, 1968, *J. Phys. C* **1,** 544.

J. Garland, 1963, *Phys. Rev. Lett.* **11,** 111, 114.

B. Geilikman, 1960, *Sov. Phys.—JETP* **37,** 635.

B. Geilikman, V. Kresin, and N. Masharov, 1975, *J. Low Temp. Phys.* **18,** 241.

V. Ginzburg and L. Landau, 1950, *Zh. Eksp. Teor. Fiz.* **20,** 1064.

V. Ginzburg, 1944, *Zh. Eksp. Teor. Fiz.* **14,** 177.

J. Goldman, 1947, *Zh. Eksp. Teor. Fiz.* **17,** 681.

C. Gorter and H. Casimir, 1934, *Physica* **1,** 306.

A. Hewish, 1968, *Sci. Amer.* **219,** 25.

D. Jerome, A. Hazaud, M. Ribault, and K. Bechgaard, 1980, *J. Phys. Lett.* **41,** L195.

B. Josephson, 1962, *Phys. Lett.* **1,** 251.

P. Kapitza, 1938, *Nature*, **141,** 75.
K. Kihlstrom, P. Movda, V. Kresin, and S. Wolf, 1988, *Phys. Rev. B* **38,** 4588.
W. Knight, 1949, *Phys. Rev.* **76,** 1259.
V. Kresin, V. Litovchenko, and A. Panasenko, 1975, *J. Chem. Phys.* **63,** 3613.
V. Kresin, H. Gutfreund, and W. A. Little, 1984, *Solid State Commun.* **51,** 339.
V. Kresin, 1987, *Phys. Lett. A* **122,** 434.
L. Landau, 1932, *Phys. Zs. Sow. Un.* **1,** 285.
L. Landau, 1941, *J. Phys. USSR* **5,** 71.
L. Landau and E. Lifshitz, 1977, Quantum Mechanics, Sec. 45, Oxford, New York.
W. A. Little, 1964, *Phys. Rev.* **156,** 396.
A. Mackintosh, 1963, *Scientific American* **209,** 110.
J. S. Martens, G. K. G. Hoenwarter, J. B. Beyer, J. E. Nordham, and D. S. Ginley, 1989, *J. Appl. Phys.* **65,** 4507.
F. Meunier, J. Burger, G. Deutcher, and E. Guyon, 1968, *Phys. Lett. A* **26,** 309.
A. Migdal, 1960, *Sov. Phys—JETP* **37,** 176.
W. McMillan, 1968, *Phys. Rev.* **175,** 537.
W. McMillan, 1968, *Phys. Rev. B* **167,** 331.
R. Miller, R. Mein, J. Gibson, J. Hulm, C. Jones, and R. Mazersky, 1965, *Proceedings of the Ninth International Conference on Low Temperature Physics,* (J. Daunt, D. Edwards, F. Milford, and M. Yaguf, eds.) Plenum, New York, p. 600.
H. K. Onnes, 1911, *Leiden Commun.* **124C**.
D. Pines, 1956, *Can. J. Phys.* **34,** 1379.
R. Simon and P. Chaikin, 1981, *Phys. Rev. B* **23,** 4463.
R. Simon and P. Chaikin, 1984, *Phys. Rev. B* **30,** 5552.
A. Sleight, J. Gillson, and P. Bierstedt, 1975, *Solid State Commun.* **17,** 27.
F. Steglich, J. Aarts, C. Bredl, W. Lieke, D. Merschede, W. Franz, and H. Schafel, 1979, *Phys. Rev. Lett.* **43,** 1892.
B. Stritzker and W. Buckel, 1972, *Z. Phys.* **257,** 1.
H. Suhl, B. Mattiass, and L. Walker, 1959, *Phys. Rev. Lett.* **3,** 552.
U. Welp, W. K. Kwok, G. W. Crabtree, K. G. Vandervoort, and J. Z. Lin, 1989, *Phys. Rev. Lett.* **62,** 1908.
S. Wolf, J. Kennedy, and M. Nisenoff, 1978, *J. Vac. Sci. Technol.* **13,** 145.

HIGH-T_c OXIDES

Reviews

J. Bednorz and K. A. Mueller, 1988, Nobel lectures in physics, *Rev. Mod. Phys.* **60,** 585.
R. Fischer, J. Gordon, and N. Phillips, 1988, Heat capacity of high T_c oxides, *J. Superconductivity* **1,** 231.
D. Ginsberg (ed.), 1990, *Physical Properties of High Temperature Superconductors,* World, Singapore.
R. McConnell and S. Wolf (eds.), 1989, *Science and Technology of Thin Film Superconductors,* Plenum, New York.
A. Narlikar (ed.), 1989, *Studies of High* T_c *Oxides,* Nova, New York.
W. Pickett, 1989, Electronic Structure of the High Temperature Oxides Superconductors, *Rev. Mod. Phys.* **61,** 433.
C. Poole, Jr., T. Datta, and M. Farach, 1988, *Cooper Oxide Superconductors,* Wiley, New York.

Papers

A. Arko, R. List, R. Bartlett, S. Cheong, Z. Fisk, J. Thompson, C. Olson, A. Yang, R. Liu, C. Gu, B. Veal, J. Liu, A. Paulikas, K. Vandervoort, M. Claus, J. Campusano, J. Schrieber, and N. Shinn, 1989, *Phys. Rev. B* **40,** 2268.

S. Barrett, D. Durand, C. Dennington, C. Slichter, T. Friedmann, J. Rice, and D. Ginzberg, 1990, *Phys. Rev.* **41,** 6283.

A. Bednorz and K. A. Mueller, 1986, *Z. Phys. B* **64,** 189.

P. Boni, J. Axe, G. Shirane, R. Birgeneau, D. Gaffe, H. Jenssen, M. Kastner, C. Peters, P. Picone, and T. Thirston, 1985, *Phys. Rev. B* **38,** 185.

I. Bozovik, 1989, *Physica C* **162–164,** 1239.

R. Cava, B. Batlogg, J. Krajewski, R. Farrow, L. Rupp, Jr., A. White, K. Short, W. Peck, and T. Kometani, 1988, *Nature* **333,** 814.

C. Cough, M. Coldough, E. Fargan, R. Jordan, M. Keene, C. Muirhead, A. Rae, N. Thomas, J. Abell, and S. Sutton, 1987, *Nature* **326,** 855.

M. Crawford, M. Kunchurt, W. Farneth, E. McCarron, and S. Poon (unpublished).

G. Deutcher, 1987, in *Novel Superconductivity,* S. Wolf and V. Kresin (eds.), Plenum, New York, p. 293.

R. Fisher, S. Kim, S. Lacy, N. Phillips, D. Morris, A. Markelz, J. Wei, and D. Ginley, 1988, *Phys. Rev. B* **38,** 11942.

D. Ginsberg, S. Inderhees, M. Salamon, N. Goldent, J. Rice, and B. Pazoe, 1988, *Physica C* **153–155,** 1082.

V. Ginzburg, 1989, *J. Superconductivity* **2,** 323.

M. Gurvitch, J. Valles, Jr., A. Cucolo, R. Dynes, J. Carno, L. Schneemeyer, and J. Waszczak, 1989, *Phys. Rev. Lett.* **63,** 1008.

Q. Huang, J. Zasadzinski, and K. Gray, preprint.

J. Imer, F. Pattheyt, B. Dardel, W. Schneider, Y. Baer,Y. Petroff, and A. Zettl, 1989, *Phys. Rev. Lett.* **62,** 336.

Y. Jean, J. Kyle, H. Nakanishi, P. Turchi, R. Howell, M. Fluss, A. Wachs, R. Meng, P. Hor, J. Huang, and C. Chu, 1988, *Phys. Rev. Lett.* **60,** 1069.

A. Jesowski, J. Mucha, K. Rogacki, P. Moryn, Z. Bukowski, M. Hovofiowski, J. Tatalowicz, J. Stepien-Damm, S. Sulkowski, E. Trojnar, A. Zaleski, and A. Klamut, 1987, *Phys. Lett. A* **122,** 431.

V. Kresin, 1987, *Phys. Rev. B* **35,** 8716; 1987, *Solid State Commun.* **51,** 339.

V. Kresin and H. Morawitz, 1988, *Phys. Rev. B* **37,** 7854; 1988, *J. Superconductivity* **1,** 108; 1990, *Phys. Lett.* **145,** 368; 1990, *Solid State Commun.* **74,** 1203.

V. Kresin and S. Wolf, 1987, *Solid State Commun.* **63,** 1141; 1988, *J. Superconductivity* **1,** 143.

V. Kresin and S. Wolf, 1990, *Phys. Rev. B* **41,** 4278.

V. Kresin, G. Deutcher, and S. Wolf, 1988, *J. Superconductivity* **1,** 327.

W. Little, 1989, *Phys. C* **161,** 195.

H. Maeda, Y. Tanaka, M. Fukutomi, and T. Asano, 1988, *Jpn. J. Appl. Phys. Lett.* **27,** 209.

D. Marshman, G. Aeppli, B. Batlogg, J. Brenner, J. Carolan, R. Cava, M. Celio, A. Chandler, W. Hardy, S. Kreitzman, G. Luke, D. Noakes, and M. Senba, 1987, *Phys. Rev. B* **36,** 2386.

A. Masaki, H. Sato, S. Uchida, K. Kitazama, S. Tanaka, and K. Inoue, 1987, *Jpn. J. Appl. Phys.* **26,** 405.

F. Mueller and J. Smith (unpublished).

K. A. Muller, preprint.

N. Ong, Z. Wang, J. Clayhold, J. Tarascon, L. Greene, and W. McKinnon, 1987, *Phys. Rev. B* **35,** 8807.

N. Phillips, R. Fisher, J. Lacy, C. Marcenat, J. Olsen, W. Ham, and A. Stacy, 1987, in *Novel Superconductivity,* S. Wolf and V. Kresin (eds.), Plenum, New York.

M. Sera, S. Kondoh, K. Fukuda, and M. Sato, 1988, *Solid State Commun.* **66,** 1101.
Z. Sheng and A. Hermann, 1988, *Nature* **332,** 55.
F. Slakey, M. Klein, E. Bukowski, and D. Ginsberg, 1990, *Phys. Rev. B* **41,** 2109.
M. Suzuki, 1989, *Phys. Rev. B* **39,** 2312.
H. Takagi, S. Uchida, and Y. Tokura, 1989, *Phys. Rev. Lett.* **62,** 1197.
S. Tewari and P. Gumber, 1990, *Phys. Rev. B* **41,** 2619.
T. Takahashi, H. Matcayama, H. Katayama-Yoshida, Y. Okabe, S. Hosoya, K. Seki, H. Fujimoto, M. Sato, and H. Jnokushi, 1988, *Nature* **334,** 691.
J. Torrance, Y. Tokura, A. Nazzal, A. Bezinge, T. Huang, and S. Pavkin, 1988, *Phys. Rev. Lett.* **61,** 1127.
W. Warren, Jr., R. Walstedt, G. Brennert, G. Espinosa, and J. Remeika, 1987, *Phys. Rev. Lett.* **59,** 1860.
M. Wu, J. Ashburn, C. Torng, P. Hor, R. Meng, L. Gao, Z. Huang, and C. Chu, 1987, *Phys. Rev. Lett.* **58,** 908.

INDEX

A-15
  compounds, 109
  crystal structure, 110
Al_2O_3, 107
Andreev's reflection, 99
Anisotropic thermal effect, 54
Antenna, 166

B-1 compounds, 112
Ba–K–Bi–O, 118, 182
Ba–Pb–Bi–O, 118
BCS theory, 17
Bechgaard salts, 113
Bi–Sr–Ca–Cu–O, 181
Bi–Sr–Cu–O, 181
Bolometers, 163
Born–Oppenheimer approximation, 116
Bose–Einstein condensation, 128

Coherence length, 22, 214
  in the N film, 98
Conductors, power, 169
Cooper theorem, 21, 197, 215
Coulomb repulsion, 18, 25
  pseudopotential, 25, 70, 78
Coupling constant, 24, 73, 79, 220
Critical current, 19–64, 141, 145, 148–151, 191
Critical field
  H_c, 10, 26, 60, 146
  field H_{c1}, 62, 146, 148–149, 192
  field H_{c2}, 63, 65, 146, 148–149, 150–153, 193, 215
Critical state model, 148–151; *see also* Critical current
Critical temperature, 4, 6, 7, 76, 141–145
Critical velocity, 129
Cu, 163
Current–voltage characteristic, 28, 38
CVD, 106

Debye approximation, 80
Debye frequency, 20
Delay lines, 165
Density of states, 30, 212
Detectors
  Josephson junction, 161–163
  video, 157, 162
Diamagnet, 141
Digital electronics
  logic, 166–167
  memory, 167
Doping, 186

Effective mass, 201, 212
Electron–phonon interaction, 24
Electronic sound, 206, 221
Eliashberg equation, 70
  function, 73
Energy gap, 18–22, 27
  anisotropy, 33
  at finite temperature, 23
Energy losses, 207
$(ET)_2X$ molecule, 113
Evaporation, 101–103
Excitons, 87

Fermi curve, 211
Fermi energy, 19, 27, 211
Fermi liquid, 203
Fermi surface, 19, 33, 35, 202
Fermi velocity, 212
Fermiology, 202
Films, 93–107
Filters, high power, 177
Fluctuations, 87, 215
Flux
  quanta, 156
  transformer, 159

Gapless state, 37
Generator, 172
Ginzburg-Landau theory, 13, 58
Granular superconductor, 95–97, 191

Hall effect, 186
Hard superconductors, 64
Heat capacity, 11, 33, 37, 47
  jump, 12, 48, 189
Heavy fermions, 89, 115
Heisenberg uncertainty relations, 122
High Q
  element, 165
  filter, 165
Hydrides, 118

I–V characteristic, 147
Ideal diamagnetism, 9
  in molecules, 139
Intervalley transitions, 117
Ion beam sputtering, 104
IR detector, 162
Isotope effect, 14, 17, 25, 81, 90, 219
  negative, 83, 119

K_2NiF_4 structure, 180
Knight shift 68, 190, 217
Kosterlitz–Thouless–Berezinski transition, 94

La–Ba–Cu–O, 179
La–Sr–Cu–O, 180, 210
La_2CuO_4, 186
$LaAlO_3$, 103
Laser evaporation, 105
Law of corresponding states, 26, 36
Little's model, 86

Localization, 97
London equation, 13
Lorentz force, 62, 64

Magnet for
  energy storage, 171
  generator, 172
  high energy physics, 170
  levitation, 173
  MRI, 172
  motor, 172
  separation, 175
Magnetic flux, 39
  quantum, 40
Magnetic impurities, 37
McMillan–Dynes equation, 77
McMillan–Rowell method, 73
Meissner effect, 8, 55, 57
MgO, 107
Microwave cavities, 66
Microwave component
  delay line, 165
  filter, 165
  mixer, 157
  SIS, 162
  switch, 166
Mixed state, 60, 189
MOCVD, 106
Molecular spectroscopy, 92
Multigap structure, 35, 216

Nb, 103, 107, 109, 111, 163
Nb_3Ge, 110, 183
Nb_3Si, 110
Nb_3Sn, 103, 146, 170
NbCN, 112
NbN, 104, 107
NbTi, 146, 169
Nd (Ce, Pr) CuO, 182
Neutron spectroscopy, 75, 91, 127
Neutron stars, 133
Nuclear magnetic resonance, 68
Nuclear momenta of inertia, 132

Ohm's law, 56
Optical detection, 162
Order parameter, 59
  energy dependent, 70
"Orsay" group, 99
Overlapping bands, 87

Oxides, 118

Patterning, 108
Peak effect, 65
Penetration depth, 10, 57, 58, 93, 163
Percolation, 95–96
Persistent current, 39
Phonon–plasmon mechanism, 221
Pinning, 64, 95
Plasmons, 205
  demons, 88
  Quasi-2D plasmons, 89, 188, 205, 214
  Plasmon band, 188, 206
Positron lifetime, 189, 216
Proximity effect, 97
  system, 59
  tunneling model, 99
  critical temperature, 100
Pulsars, 134

Quantum fluid, 121
Quasiparticles, 22, 47, 124, 199

Renormalization function, 70
Resistance measurement, 141–143

S-wave, 190
Skin depth, 66
SNS, 194
Sommerfeld constant, 186, 203, 218
Spin-orbit interaction, 68
Sputtering, 103–104
  ion beam, 104
SQUID
  dc, 157–160, 194
  rf, 160–161
$SrTiO_3$, 108
Starquake, 136
Strong coupling, 26, 48
Substrates, 107
Superconductivity of Type I, 59, 61
Superconductivity of Type II, 59, 62
Superfluidity, 123
Surface energy, 61
Surface impedance, 67
Surface resistance, 163
Susceptibility
  ac, 143–144
  dc, 144
  magnetic, 141

Ta, 109
Thermal conductivity, 12, 48, 186, 190, 213
  electronic component, 48, 52
  lattice component, 48, 50
  types, 53
Thermal phonons, 80
Tl–Ba–Ca–Cu–O, 181
TMTSF molecule, 113
Transformer, dc, 95
Transmission lines, 66
  power, 175
  signal, 165
Tunneling, 27, 190
  matrix element, 29
  junction, 30, 31, 73
  density of states, 73
Two-dimensional potential, 197
Two-fluid model, 12, 13, 23, 24, 38, 51, 66
  for He II, 120
  for nuclei, 133

Ultrasonic attenuation, 32, 34
  absorption coefficient, 32

Va, 103, 109, 111
Va_3Ga, 103, 111, 170
Van der Waals forces, 131
Variable electromagnetic field, 65
Vortex pinning, 146
Vortices, 63, 93, 146
  in He II, 129
  in neutron stars, 135

Wiedemann–Franz law, 50

$Y_1Ba_2Cu_3O_7$, 103, 108, 181, 217

Zero-point vibrations, 122, 199
Zr, 109

9 780306 434747